D0214567

Human Origins 101

Recent Titles in the
Science 101 Series

Evolution 101
Janice Moore and Randy Moore

Biotechnology 101
Brian Robert Shmaefsky

Cosmology 101
Kristine M. Larsen

Genetics 101
Michael Windelspecht

Human Origins 101

Holly M. Dunsworth

Science 101

GREENWOOD PRESS
Westport, Connecticut • London

Library of Congress Cataloging-in-Publication Data

Dunsworth, Holly M.
Human origins 101 / Holly M. Dunsworth.
p. cm. — (Science 101, ISSN 1931-3950)
Includes bibliographical references and index.
ISBN 978–0–313–33673–7 (alk. paper)
1. Human beings—Origin. 2. Human evolution. I. Title.
GN281.D865 2007
599.93′8—dc22 2007022747

British Library Cataloguing in Publication Data is available.

Library of Congress Catalog Card Number: 2007022747
ISBN-13: 978–0–313–33673–7
ISSN: 1931-3950

First published in 2007

Greenwood Press, 88 Post Road West, Westport, CT 06881
An imprint of Greenwood Publishing Group, Inc.
www.greenwood.com

Printed in the United States of America

The paper used in this book complies with the Permanent Paper Standard issued by the National Information Standards Organization (Z39.48–1984).

10 9 8 7 6 5 4 3 2 1

For Mom and Dad,
Model Human Beings

Contents

Series Foreword

What should you know about science? Because science is so central to life in the 21st century, science educators believe that it is essential that *everyone* understand the basic foundations of the most vital and far-reaching scientific disciplines. *Human Origins 101* helps you reach that goal—this series provides readers of all abilities with an accessible summary of the ideas, people, and impacts of major fields of scientific research. The volumes in the series provide readers—whether students new to the science or just interested members of the lay public—with the essentials of a science using a minimum of jargon and mathematics. In each volume, more complicated ideas build upon simpler ones, and concepts are discussed in short, concise segments that make them more easily understood. In addition, each volume provides an easy-to-use glossary and an annotated bibliography of the most useful and accessible print and electronic resources that are currently available.

PREFACE

The short answer to the question of human origins has already been triangulated by genetics, paleontology, and archaeology: 200,000 years ago in sub-Saharan Africa.

So why do we need an entire book just to introduce the subject? Because the long answer is an ever-lengthening saga that contains as many chapters as there are traits that make up the human species. Different aspects of *Homo sapiens* arose at different times to build the creatures we are today. The three tiny bones in our ears for hearing evolved over 100 million years ago, but we did not begin to make musical instruments until at least 100,000 years ago. Each acquisition of a human trait affected events further along the human evolutionary path. It is because of this broader quest for human origins that there is more to discuss than simply "200,000 years ago in sub-Saharan Africa."

If every human drew their family tree all the way back to the beginning of life on planet Earth, each person's history would be identical from the Big Bang until very recently, when our individual histories—the twigs on our tribal lineages—diverged from the branches of other human groups and became geographically and genetically distinct. Just like a person can trace her curly hair back through generations by looking at photographs of her cousins, grandparents, and great-grandparents, the human species can trace its roots back by looking at fossils and genes that reveal our shared ancestries with our extinct hominin ancestors and with our "cousins"—apes, monkeys, mice, horses, fishes, worms, corn, slime molds, bacteria, and every living thing that humans have ever bothered to name.

It is because of our intellect that we often very easily forget that we are members of the animal kingdom, working under and shaped by the same basic evolutionary processes as all other living things. Despite the long list of differences that we imagine separate or elevate us from other animals, we differ from chimpanzees by less than 2 percent of our genetic code (which is not much considering it is 3 billion base-pairs long). Certainly it is our differences from other animals that define us as humans, but it is our fundamental similarities to those animals that eventually allowed us to become humans.

Knowing we are upright, chattering African apes need not send us into an existential crisis, not with the knowledge that each one of us is unique. No two people have exactly the same DNA, not even adult twins. No one in the past or future will ever have the same genetic code as any other person in the past or future, or any other organism for that matter. Each one of us is the result of a single successful, uninterrupted chain of life that began 3 billion years ago. Instead of using this strangely sharp intelligence to wallow in our ordinary primateness, we should, instead, marvel at what natural selection and other biological forces produced from an ancestor with monkey-like brains.

As arguably the only self-conscious creatures on the planet, it is shocking how little we know about ourselves. Many people, if asked, will tell you they are a "Homo sapien," incorrectly assuming the species in our scientific name is plural. Some people could assemble the parts of a sports car down to the last ball bearing, but could not locate their own semi-circular canals, let alone describe what they do. (They are in the ears but they have nothing to do with hearing. They are involved in seeing clearly while the head is moving around.)

Through oral and written histories we immortalize our family genealogies, we even name our babies after their ancestors, but we can rarely be bothered to remember the very long, strangely pronounced names of our extinct evolutionary ancestors. Next to the anatomy books on our library shelves are volumes instructing us in the basic aspects of life, things other animals do instinctually or learn from one another without spoken or written language, like reproducing, raising offspring, making friends, running efficiently, hunting effectively, eating the right foods, and surviving in "the wild" without a mobile phone.

In the following pages, the overwhelming evidence for human origins and evolution and the fundamental concepts used to interpret that evidence are introduced and discussed as only humans can do, with an arbitrary system of symbols printed on processed tree pulp. Reading about the first spear throwers, prehistoric mammoth-hunting injuries, or the first sea-faring journey to Australia should be like reading about the miraculous throwing arm on your great-uncle, about how your great-grandfather died reeling in a 150-lb fish, or about how your great-great-great grandmother emigrated to America on a boat from Ireland.

It is my hope that readers of *Human Origins 101* take away this important tenet of human origins studies: we are modified African apes that, despite seemingly great variation in biological and cultural ornamentation, share a common African ancestor. We are all one diverse species with regionally varying physical characteristics that are the results of environmental adaptations, mate preferences, migration, or simply chance. Culture has a significant effect on the differences we perceive in other people. Underneath the t-shirts, face paint, piercings, tattoos, mohawks, and stilettos, we are all remarkably similar. It is because of our unique makeup that we have both the propensity to forget and the ability to embrace that we are an integral part of the natural world around us.

Acknowledgments

For inspiration, discussion, and guidance, I am happily indebted to Kevin Stacey, Julie Dunsworth, Cheryl Hill, Pat Shipman, David Carlson, Brian Regal, members of the Penn State biological anthropology journal club, and the students from my spring 2006 course in biological anthropology. Alan Walker and Jeffrey Kurland kindly allowed me to bask in the glow of their very clever brains for the last seven years. Enormous gratitude goes to Greenwood editor Kevin Downing. Drawings and maps were artistically prepared by Jeff Dixon. Thanks to Kristina Aldridge, Mark Teaford, David Lordkipanidze, Chris Campisano, Alice Teeple, and students of the 1998 Rutgers Koobi Fora field school for assisting with photographs and figures. Any errors on the following pages are mine alone. Few of the ideas presented in these pages are new ones and I thank all of the people responsible for them. Thanks most of all to the people who work in hot deserts, cold caves, lonely museum stacks, and sterile laboratories for their tireless curiosity and enthusiasm for human origins and evolution.

Introduction

Piecing together the puzzle of human origins and evolution is an interdisciplinary endeavor. The fields of genetics, paleontology, paleoanthropology, archaeology, cultural anthropology, primatology, animal behavior and ecology, anatomy, physiology, kinesiology, psychology, cognitive sciences, economics, and many others offer pieces of the puzzle.

The roots of humankind can be traced back to the beginnings of the universe because we are composed of atoms, to the dawn of life on earth because we are carbon-based, to the first fish to make a living on land because we are tetrapods, to the small mammals that survived the dinosaur-ending apocalypse because we suckle milk as babies and are furry (some more than others), to the apes that for some strange reason got good at teetering around on their hind legs, and finally to expert bipeds with big brains for mastering tool-making and language.

Certainly the entire history of the universe as it pertains to human origins and evolution cannot be covered in this volume. The main focus of human evolutionary studies (and the main focus here) is the roughly 6 million crucial years from the moment our lineage split from the ancestors we share with chimpanzees up until the first modern humans emerged in Africa, just before our individual and population histories began to diverge.

Although most technical terms are defined in the glossary, there are two that need special attention from the start. *Hominins* are modified African apes. They include living humans and all the extinct descendents of the last common ancestor with chimpanzees that are on the human lineage as opposed to the chimpanzee lineage. Fossil hominins are either our direct ancestors or are our evolutionary cousins located on different branches of the hominin tree from ours. Most hominins are bipedal and relatively large-brained, but the very earliest ones may not have been so, if those traits did not evolve immediately after the split from the common ancestor with chimpanzees. *Humans* are the only surviving hominin species and are distinct from other hominin species in our unique combination of complex material culture, social behavior, bodily characteristics, and intelligence. In this book, we will refer only to anatomically modern humans or *Homo sapiens* as "humans." These are people we would call "people" if we were to stand eye-to-eye with one of them. To put it another way, if we traveled back in a time machine to 200,000 years ago we would probably recognize the upright walking beings as human beings (even though they probably did not yet use language or symbolism like we do), but there is no guarantee that we would connect the same way with anything living before that, not necessarily even with Neanderthals who lived side by side with some human populations until 30,000 years ago. Although some scientists refer to everything within the human genus (*Homo*) as "humans," we will reserve that title for ourselves only.

It seems logical that if one is to study human origins there should be universal agreement as to what exactly makes a human "human." But if asked, "what defines humans or what makes them unique?" many popular answers lie within the intellectual and emotional realm: love, laughing, regret and loss, creativity, curiosity, hunger for learning, perception of things unseen, concept of the future, spirituality, suicide, and language. These are not the types of traits that are readily traced through prehistory with fossils, artifacts, or genetics. Comparative observations of other animals help us understand the nature of these human attributes, but the more we learn, from apes in particular, the more we find that very little of what defines humans is exclusively human.

The 2 percent of our genome that differs from chimpanzees sometimes seems too small to contain all of our differences. Physically, we have a larger brain with thick hair covering the skull and, in males, the face as well. Our body hair is drastically reduced and our naked skin contains many more sweat glands. Our hands are much more dexterous but our bodies are much weaker than chimpanzees'. We have large conspicuous sex organs and we walk on only two very long legs shod with rigid, sturdy feet. Much of the white of our eyes, which is whiter than other apes, is visible due to the almond shape of our eyelids. Behaviorally, we depend on highly complex vocalizations. We control fire. We graffiti nearly everything, including ourselves. On the whole, we peacefully gather in enormous groups containing mostly unrelated individuals. We exploit and selectively breed other species (and even members of our own, in arranged marriages). We are skillful hunters without fangs and claws and are capable of adapting to every sort of environment on earth.

In the following pages we track the origin and evolution of the traits that enabled humans to become "human." If these traits will someday be mapped in the genome, we will find that they are expressed or regulated by that less than 2 percent DNA difference that separates us from chimpanzees. As we learn more about our own genome and that of related animals, we are finding that most of what we consider unique to our species probably emerged recently and because most traits are complex and expressed through multiple genes, uniquely human traits are going to be a challenge to pinpoint at the molecular level, let alone trace through evolution.

By no means can the entire field of human origins fit into a single book, so the fundamentals and the highlights of the material evidence are included here. For anatomical vocabulary as well as the epochs, periods, and eras of geologic time, please consult Appendices A and B. Throughout the text "Mya" and "Kya" will be shorthand for millions of years ago and thousands of years ago. For additional, in-depth information, check the recommended resources in Appendix C, and the references in the bibliography for places to start your own quest for human origins.

Chapter 1 begins the book with a brief history of the search for human origins and human evolutionary studies. As a consequence of self-consciousness, humans have probably been curious about their origins since the first hominin developed the mental capacity to do so. But the modern study of human origins did not seriously begin until the middle of the 19th century when the first Neanderthals were found in Europe. Early in the history of the field, the controversy surrounding new discoveries was due to uncertainty as to whether or not humans evolved at all. Now that the old controversy has been conquered by science, controversy surrounding new discoveries merely refers to the details involved in their interpretation. Unless fossils come with identification tags, there will always be controversy about how to interpret them. For most every issue, however, there is a majority of scientists who agree on an interpretation which becomes the working paradigm unless a new discovery flips everything upside down again. This collegial debate over new finds is still, unfortunately, repackaged and then touted by antievolutionists as proof that there are flaws in evolutionary theory.

Because this book is centered on the processes of evolution, *Chapter 2* outlines evolutionary theory and walks through Darwin's formulation of the theory of evolution by natural selection and also by sexual selection.

It focuses on how we understand evolution today with our current grasp of genetics. In this chapter we look to the evidence we glean from the living world around us for human origins and evolution, so methods of classification and forming family trees are discussed and then groups within the Order Primates, of which we are a member, are discussed.

Without time travel, one can never be truly certain what life was like before moving pictures were first recorded or before people began to write down history. Old bones and stones, however, provide an astonishing amount of information about the origins and evolution of humans. In *Chapter 3*, we dig down to bare bones of our search for evidence of human origins, into the fossil and archaeological records. We meet our ancestors and our close cousins, and we sort through the garbage they left behind. We discuss the fossils and artifacts that by chance got preserved, that by hard work and a little luck got discovered, and that by careful scientific scrutiny got interpreted as evidence for the nature of our origin and evolution. Our body's architecture and mechanics are only part of the vastly underappreciated knowledge researchers are compiling everyday on our evolution and place in nature. The establishment of geologic methods in the beginning of the chapter puts the fossils into perspective. Early fossil primates, monkeys, and apes are briefly toured and then once hominins are reached they are organized into subheadings by species. Only the general anatomical features are discussed, but keep in mind that scientists base species distinctions and anatomical interpretations on careful measurements and statistical analyses.

Because so much of evolutionary theory relies on modern genetics, *Chapter 4* includes advances that the field of genetics has made in our understanding of human origins and evolution. Through DNA, human ancestry can be traced as far back as the origins of life and multicellular organisms. With increasingly powerful biotechnology, the search for human origins is no longer simply based on dusty fossils and artifacts. Today, artifacts within the *genome* are just as important as those buried in the ground. In an age where molecular analysis is ever increasing in precision and scope, we can use it as a tool for tracking our evolution. Knowledge of genes and inheritance from Chapter 2 is used to explore the use of molecular clocks to determine lineage-splitting times in prehistory, the role of mitochondrial DNA in determining our African ancestry, and also the evolution of some recent human adaptations. Although ancient DNA is not technically modern evidence, the techniques by which it is extracted and analyzed are modern, so it is included here.

In *Chapter 5* we synthesize the evidence from living animal models, fossils, artifacts, and genetics in order to track the origins and evolution of some fundamental human traits. Consider this chapter a blueprint for building a human from an ape-like ancestor as opposed to building a chimpanzee from that same ancestor. The evolution of many, if not all, human traits are highly dependent on the evolution of others. For example, the adoption of habitual bipedalism allowed selection to act on the anatomy of newly freed hands that became even more adept at using and making increasingly elaborate tools. Therefore, one cannot adequately study bipedalism without understanding the archaeological record of stone tools.

Our ancestors were shaped by the same natural processes and evolutionary forces as all other earthly organisms. But despite worldwide dispersal, cultural complexity, the innovative modifications we make to our surroundings, and the great impact we have on the environment, are those same forces working on humans today? What spurred modern humans to spread across the world in the first place and how and when did world colonization occur? Will we have time to evolve much more or will we go extinct before we can find out? And if given the chance to do it all over again, would humans still evolve? These questions are asked in *Chapter 6*.

Evolutionary research is revolutionary. With additional discoveries from dig sites and from the use of new technologies, not only can hypotheses about human origins and evolution be tested and retested from many different scientific perspectives, but they are constantly tweaked to reflect the influx of new evidence. Tomorrow someone could find a new piece of the puzzle that either reinforces what most people already knew or that totally overturns their perspectives and forces them to rethink the whole thing. A new fossil species or a genetic breakthrough can change the way we think about our origins and evolution literally overnight.

If, hypothetically, someone tomorrow discovers a fossil hominin from 9 Mya that walked upright, then parts of this book will need to be rewritten (because as it stands now, the overwhelming evidence points to a hominin origin at about 6 Mya). But no matter the future of the science of human origins and evolution, the information contained in this volume serves as a basis for understanding why such a discovery would be significant and how it would impact the current understanding based on the overwhelming evidence from the fossil, archaeological, and genetic records.

1

A Brief Overview of the Search for Human Origins

THE SCIENCE OF HUMAN ORIGINS AND EVOLUTION

Whether they study Ice Age cave paintings, chimpanzee DNA, or the bones of the first tiny squirrel-like primates from 60 million years ago, scientists are asking the same questions: Where did we come from and how did we get here?

Although the search for human origins draws upon research from many scientific disciplines, it is mainly kept to the field of anthropology. Broadly defined as "the study of humans," anthropology can encompass any scientific pursuit as long as it has a human focus.

Biological anthropology (also called physical anthropology) is a field within anthropology, and all of its subfields contribute directly or indirectly to the understanding of human origins and evolution. These subfields include scientific investigations of the genetics, behavior, biology, ecology, and evolution of humans and nonhuman primates that fall under disciplines like "paleoanthropology," which is the study of human evolution (Figure 1.1), and "primatology," which is the study of living primate behavior and ecology.

Biological anthropology does not stand alone in the search for human origins. The other anthropological fields of archaeology and cultural anthropology are also crucial to the understanding of human origins and evolution.

Because paleoanthropology reconstructs the ancient past, it is a historical science. Thus, it is difficult to make the same types of conclusions that chemists or cell biologists can make from eye-witnessing experiments. In paleoanthropology, "Mother Nature" has performed the experiments and thousands or millions of years later, scientists describe

Figure 1.1 A group of young paleoanthropologists excavate an early Pleistocene site at Koobi Fora, Kenya. By systematically digging even layers, they have taken down the hillside to find bones and stone tools. The sediment is placed into bowls and then carried to a sieving station where it is shaken through a fine mesh so that even the tiniest teeth, bones, and even fossil seeds can be recovered. The wall on the side of the excavation will be useful for viewing the ancient sediments and for determining how the site was formed and how the artifacts were buried. *Photograph by Holly Dunsworth.*

the ancient experiments sight-unseen and explain the results based on the evidence that survived.

SCIENTIFIC METHOD

When a paleoanthropologist finds a fossil (Figure 1.2), she applies the scientific method to decipher its place in evolutionary history. Based on prior knowledge gained from other scientists' work and from her own observations, she forms a hypothesis. Then she collects evidence, or data, to test her hypothesis. This is the step where a chemist would perform an experiment, but a paleoanthropologist, instead, collects evidence of evolution's experiment by analyzing other fossils and the bones of living species that are similar.

If the data she collects supports her hypothesis, then the paleoanthropologist and other researchers repeat the scientific process to determine if additional evidence can support the same hypothesis. If it does, and

Figure 1.2 This photograph of the right side of a mandible, or lower jaw, of the extinct stem ape *Proconsul* was taken moments after its discovery. It was found peeking out of the ground at site R106 on Rusinga Island, Kenya, in the summer of 2006. For orientation, the molar teeth are on the left and they are heavily worn away from poor preservation. *Photograph by Holly Dunsworth.*

if the hypothesis continues to be supported, then that hypothesis turns into a theory. The stronger the support is for the theory, the stronger the evidence against it must be to falsify it. As it gains support in the form of repeated testing that theory will eventually become so strong that it will become part of the prior knowledge that other scientists apply to the formation of new hypotheses. The cycle continues endlessly with hypotheses being tested and either supported or not, with theories being overturned by the discovery of evidence that refutes them, with entirely new hypotheses being born when theories crumble, and with some theories standing the test of time.

Before collecting any fossils, a paleoanthropologist selects a field site based on the types of fossils that she can use to test her hypotheses. So, if her hypotheses are centered on the split between the chimpanzee and human lineages, she surveys a site with a known geologic age or one that corresponds in age to other sites that have already produced very early hominin fossils. She searches for rocks that match the age of genetic time estimates for the split around 6 Mya (see discussion of "Molecular Clocks" in Chapter 4).

Along with any fossils of chimpanzee or human ancestors, the paleoanthropologist must also observe and collect the other types of fossils at the site. The preserved plant and animal remains will help reconstruct the ancient environment. The nature of the rock layers will also indicate whether a lake, a delta, a river, a gully, an animal burrow, a den, or many other types of burial scenarios deposited the sediment. (Hominins are not found inside intentionally dug graves until very recently.) These types of environmental reconstructions will put the species into an environmental context and will also provide additional information about ape and human evolution that the fossil cannot provide itself.

In the laboratory, the paleoanthropologist will analyze the anatomy of the fossils by recording measurements (e.g., the lengths and widths of the bones and teeth) and by looking inside the bones with imaging technology like x-ray computed tomography (i.e., "CT" or "cat" scanning; (Figure 1.3)). These data are then compared with other fossils and to skeletons of modern apes and humans.

The paleoanthropologist will test the hypothesis that the fossil she collected is more closely related to chimpanzees or to humans by testing if the anatomical traits are more like one than the other. If the fossil's characteristics are more like modern chimpanzees than humans, then it will be placed in the evolutionary lineage of chimpanzees. This placement is a hypothesis that can be overturned with additional evidence. After all, it is possible that an ape-like creature living 6 Mya was neither an ancestor of chimpanzees or humans because it may have belonged to a lineage that went extinct and did not contribute to our evolution or to that of our ape cousins.

The overarching framework in the quest for human origins is a theory that has achieved factual status: evolution by natural selection. The scientifically rigorous investigation of human origins and evolution was able to flourish after Charles Darwin made this contribution, partly because he provided hypotheses to test. For instance, in *The Descent of Man* (1871), Darwin postulated that fossils of the last common ancestor (LCA) of great apes and humans would be found in Africa since gorillas, bonobos, and chimpanzees, which are the most humanlike animals, currently live there. According to the current fossil record, Darwin's prediction is still correct, and, of course, he could be proved wrong if scientists find the evidence. It is unlikely, but possible nonetheless, that

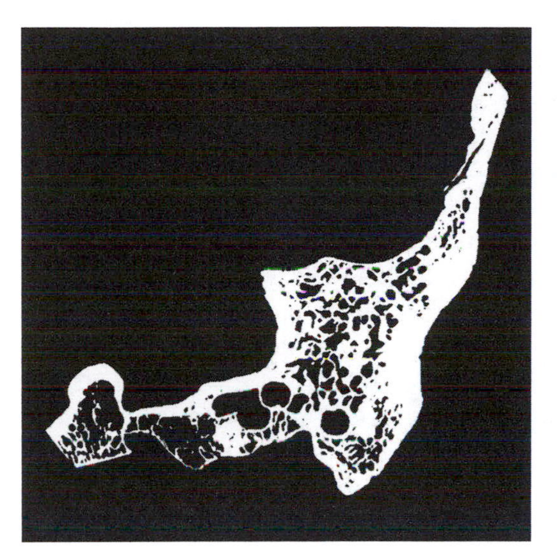

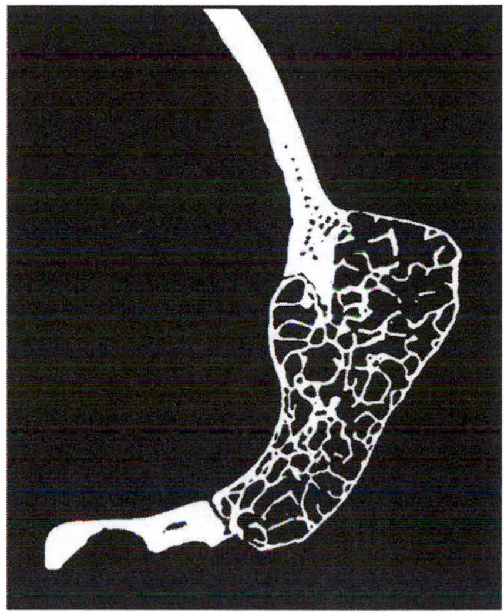

Figure 1.3 A high-resolution virtual slice through a human skull (left) and a chimpanzee skull (right) at the temporal bone near the ear. The sponge-like bone is the inside of the mastoid process which is the palpable bump just behind the ear that attaches muscles from the collarbone to the head. These are not fossilized, but many times fossils will still contain much of the infrastructure of bone, like that seen here, despite their transformation into rocks. Imaging techniques using x-ray computed tomography (also called "CT" or "cat" scanning) can reveal the inner structure of fossils-detailing their bony beginnings that distinguish them from ordinary rocks-and any anatomical clues that can be linked to behavior or can help place the fossil into an evolutionary context. It is not clear what function these air cells serve in the skull's temporal bone but they are clearly different between humans and chimpanzees and will help diagnose whether fossils that preserve these air cells are more closely related to humans or to chimpanzees. *Image courtesy of Cheryl Hill.*

tomorrow someone could dig up the LCA in Asia and turn everything upside down. That is the fundamental nature of paleoanthropology and of science.

FOREFATHERS

Aside from the epic role of Mother Nature, there are no known *foremothers* who contributed directly to human origins and evolutionary science prior to the 20th century. Furthermore, much of the theoretical and philosophical background of the science of human origins and evolution is based in Europe because that is where science as we know it was born. But for Europeans and non-Europeans alike, the quest to discover human origins and to understand human evolution resonates deeply for everyone who has ever wondered about human nature and how humans came to exist. Fossils pique our curiosity and written records, folklore, and the collection of artifacts indicate that people have always been fascinated with the things that crumble out of the earth.

Chinese folk healers still follow the ancient belief that "dragon bones"—the fossils of dinosaurs and other animals—are effective medicines. Decorations on pottery and written records reveal that, for the Classical Greeks and Romans, fossilized bones from the ground evoked questions and helped support beliefs about their own origins. Prominent families traced their pedigrees back to heroes of mythological importance. Fossils of large extinct animals, like ancient elephants, from the Miocene and Pleistocene reaffirmed their beliefs that giants and larger-than-life heroes of myths preceded them.

Today museums around the world display fossils from their particular region and boast that their part of the world is crucial to human origins. This ancient practice was invented by the Emperor Augustus (63 BC–AD 14), who established the world's first paleontological museum. It contained bones of giants and weapons of ancient heroes, all used as propaganda to empower his emerging Roman empire.

But before the first paleontologists and their museums, Plato and Aristotle (6th and 5th centuries BC) had already established the first recorded philosophy on human nature and human origins. They considered humans to be part of the natural world just like other organisms and put them at the top of the "Great Chain of Being." With the spread of Christianity, the chain was adopted into a ladder with God at the top, angels just below him, humans split into ranked racial categories below angels, nonhuman primates under them, mammals, then reptiles, amphibians, and fish, with plants rooted at the bottom, on top of inanimate objects like rocks.

Christian theology dominated the Middle Ages of Western Europe and the world was seen as the product of God's plan. Species were considered fixed and immutable and to appear today as they had always been. Humans were thought to be separate from the natural world around them and were created by God in their present form with language and culture. Biblical creation was assumed to have occurred very recently because of calculations by scholars like Bishop Ussher. He summed up ages of generations of people in the Old Testament and calculated that the creation of the world, recounted in the book of Genesis, occurred in 4004 BC

The scientific revolution of the 17th century was ushered in by Copernicus and Galileo who realized that the world revolves around the sun and not the other way around. The long-held notion that the Earth is flat was also abandoned. Furthermore, explorers increasingly came into contact with people across the oceans and were introduced to their unique cultural traditions and non-Western ways of being humans. The rigid Biblical interpretation of natural history had begun to cause problems for scientists interested in explaining the world around them.

Carl von Linné, better known by his Latinized name Carolus Linnaeus (1707–1778), took a major step toward the modern state of human origins science when he gave humans a genus and species name just like the he gave reptiles and oak trees. Linnaeus created the binomial method of classifying organisms and called humans *Homo* (man) *sapiens* (wise). The Linnaean system of taxonomy is universal today.

Georges Cuvier (1769–1832) was the first to apply scientific principles (established by Francis Bacon in 1620) to paleontology, specifically to vertebrate paleontology. He realized that catastrophes, like those described in the Old Testament of the Bible, were probably responsible for the extinctions of many of the species that are found as fossils. In Cuvier's theory of "catastrophism," areas that experience mass extinctions are restocked with new forms migrating from neighboring regions.

However, Cuvier never conceived of slow change over deep time, and that was why Charles Lyell's (1797–1875) contribution of "uniformitarianism" made him the founder of modern geology. Lyell's theory, which stands today, stated that the same fundamental natural processes that happen today (both catastrophic and slow, gradual change) have always been happening in the past and will continue to happen in the future. Thus, the Earth today looks remarkably different from how it appeared millions of years ago because of processes like erosion and volcanism that shape mountains and gorges. Lyell was one of the first to argue that the earth is indeed very old and could not possibly be the mere 6,000 years old as calculated by Bishop Ussher. Darwin read Lyell's *Principles of Geology* (1830) on his historic voyage on the H.M.S. *Beagle*, and Lyell's emphasis on the ancient age of the earth, deep time, and the gradual, slow nature of geologic processes greatly influenced him.

Before Charles Darwin formed his theory of evolution by natural selection in 1859 with *The Origin of Species,* fossils of Neanderthals and Cro-Magnons had already made giant waves in the scientific community. In 1829, the first Neanderthal remains were discovered at Engis, Belgium, but the child's skull was not recognized as a nonhuman until later. In 1848, at Forbes' Quarry near Gibraltar, an adult Neanderthal cranium was discovered and was also not recognized at the time. By 1856, an entire Neanderthal skeleton was discovered at the Feldhofer quarry in the Neander Valley, Germany. The Neanderthal as a separate, nonhuman creature—mostly separated by differences in face and skull shape as well as skeletal strength and robusticity—was established in 1864 and named according to its home region: *Homo neanderthalensis.* Fossils of modern humans had not even been known by the time Neanderthals were settled as prehistoric "cavemen." Finally by 1868 skeletons of fossilized modern humans were discovered at the Cro-Magnon rock-shelter at Les Eyzies in the Dordogne, France, hence the origin of the nickname "Cro-Magnons" for Ice Age or Late Pleistocene modern humans from Europe. The first Cro-Magnons known to science were barely younger than the Neanderthals but had clearly human traits and none of the primitive features seen on the Neanderthal skull and skeleton.

Quite the opposite of the ancient Greeks who erred frequently on the side of interpreting nonhuman fossils to be their ancestors, many of the first scientists to interpret the earliest Neanderthal and *Homo erectus* remains found in the mid to late 19th century refused to accept their role in human evolution. The so-called logical explanations usually offered for oddly shaped fossils were that they were diseased modern human remains or injured soldiers' remains, rather than they were ancient or extinct humanoids. Conversely, the scientists who did accept the paleontological evidence for human evolution became the very first paleoanthropologists. What's more, these men were no longer accidental tourists of the discipline. That is, they were no longer restricted to contemplating the incredible finds made just outside their hospital or country cottage doors. These men were scholars in human anatomy and set out with the explicit purpose to find ancient human fossils or with the explicit purpose of finding evidence for human evolution.

For instance, Thomas Henry Huxley (1825–1895) wrote a book with a title that today might be misinterpreted as an outdoor recreation manual rather than an evolutionary epic: *Evidence as to Man's Place in Nature* (1863). Huxley was the first scientist to explicitly compare the anatomy of humans and great apes, working under what became his famous charge:

> The question of questions for mankind—the problem which underlies all others, and is more deeply interesting than any other—is the ascertainment of the place which Man occupies in nature and of his relations to the universe of things. (T.H. Huxley, 1863)

Huxley argued that humans and the African great apes are very closely related, and, in fact, more so than they are to the Asian apes. He hypothesized, therefore, that human ancestry was likely to be found in Africa—a prediction that Darwin later supported in his book *The Descent of Man* (1871) and that future paleoanthropologists confirmed as well.

PILTDOWN

Early evolutionary scholars, including Darwin, anticipated that fossils of our earliest ancestors would be ape-like. However, before the fossil record was as rich as it is today, it was widely assumed that hominins evolved large brains before they evolved humanlike bodies. It was as if hominins decided with their keen intellects to walk upright (as if "up" is "right") and to use tools with their freed hands. Because of the evidence on record now, we know that bipedalism first arose about 6 Mya in tiny-brained hominins that did not make stone tools and the major expansion of the hominin brain did not occur until after 2 Mya.

This early hypothesis about brain size and its fervent following, coupled by a willingness to believe that the first brainy humans originated in Great Britain, made the notorious fossil specimen known as "Piltdown Man" such a powerful hoax.

Between1908 and 1912, a series of paleontological and archaeological discoveries were made near the village of Piltdown in Sussex, southern England. Among the remains of long extinct mammals like hippos and rhinos was a humanlike piece of skull bone and an ape-like fragment of a jaw that Charles Dawson (an amateur antiquarian and paleontologist of considerable repute) named "*Eoanthropus dawsoni.*" Many scholars were happy to accept Piltdown Man (as it was nicknamed) because it showed that big brains evolved early as a cornerstone of human evolution and that big-brained humans evolved first in England.

There was never full scientific agreement about how to fit Piltdown Man into the hominin family tree or about how to reconstruct the anatomy because there was a lack of comparative material at the time. Even when additional hominin fossils were discovered in China and Africa in the 1920s and 1930s they lacked preservation of the same parts for comparison and, furthermore, more and more hominin fossils showed that humanlike teeth evolved early while human brain size evolved much later. With the accumulation of evidence elsewhere, the Piltdown anomaly became frustrating and embarrassing to science.

With the advent and application of new technologies in the 1940s, the Piltdown problem began to unravel like a mystery. Fluorine analysis proved the skull bone and the jawbone were not from the same individual. After an animal dies the nitrogen in its bones is replaced by fluorine that is absorbed through groundwater. The levels of fluorine would have been the same in the two bones if they had been from the same animal, but they were different. Furthermore, uranium dating analyses showed that the bones were younger than 50,000 years old, which is much too recent for an ape-man.

In 1953 Joseph Weiner, Kenneth Oakley, and Wilfred Le Gros Clark published "The solution of the Piltdown problem" in the Bulletin of the British Museum of Natural History. Their analysis showed that the Piltdown jaw was from a young female orangutan and the cranium had belonged to a modern human. Piltdown was no more than a fake made to appear ancient with an artistic application of staining chemicals. Microscopic analysis showed that the teeth had been filed down to appear more ambiguous and human. The part of the jaw that articulates with the skull had been broken off to disguise that the two could not possibly fit together.

Although a hoax like Piltdown Man could not happen in this day and age, the culprit of the Piltdown forgery is still unknown. Charles Dawson is probably the best guess since he was responsible for the various discoveries at Piltdown and he was also associated with incidences of artifact fraud at other sites. However, the list of suspects continues to grow and includes author Sir Arthur Conan Doyle who participated in the Piltdown excavations, as well as Martin Hinton, a former fossil curator at the British Museum (Natural History). Recently discovered leftover laboratory materials tell of Hinton's preoccupation with the geologic processes that tarnish and stain fossils, and he could have been Dawson's accomplice in the forgery.

DUBOIS AND BEYOND

In the late 19th and early 20th centuries, there were plenty of scientists who held to the notion that the cradle of humankind would be found in Africa. But some, like Ernest Haeckel and Eugène Dubois, saw similarities between humans and orangutans of Indonesia and anticipated that human origins occurred there. Eugène Dubois was an extraordinary character in the history of human origins science. He packed up his family and his life, and moved to the Dutch East Indies on, what seemed like to his colleagues, a fanatical hunch that he would discover the "missing link" there. But in 1891 he proved them all wrong by discovering hominin fossils on the island of Java. After three years of analysis, he dubbed his find *Pithecanthropus*, which combines the Greek for both ape (pithekos) and man (anthropos). His find would later be renamed *Homo erectus*, a species which had a humanlike body but a smaller more ape-like brain. Many subsequent expeditions beginning in the 1920s recovered many more remains of *H. erectus* from other Indonesian sites as well as the famous "Peking Man" site of Zhoukoudian, China, that has produced over thirty individuals between 500 and 250 Kya.

Much more ape-like human ancestors than Dubois' fossils were unheard of until 1924 when Raymond Dart analyzed a small skull with a fossilized brain, or endocast, known as the "Taung Child" from Sterkfontein Cave, South Africa. The specimen was more primitive than *H. erectus* and supported the African origins hypothesis. He named the remains *Australopithecus africanus*, meaning "southern ape man from Africa." Critics as well as supporters of Piltdown Man's place in human evolution argued that the skull was much too ape-like to belong to the human lineage. But, Dart showed that even though it did not have identifiable features that an adult would offer, the Taung Child had humanlike traits in the teeth and skull that differed from apes.

By 1959, Robert Broom had already recognized robust australopith fossils (today called the genus *"Paranthropus"*) from South African cave sites, but a nearly complete skull found by Louis and Mary Leakey in 1959 made a much bigger splash. The Leakey's discovery of OH 5 from Olduvai Gorge, Tanzania, showed little sign of deformation and had a complete face. Immediately it was nicknamed "Nutcracker Man" and this find, after nearly twenty years of searching for human fossils led to a cascade of discoveries of more hominin fossils and stone tools from Olduvai Gorge including many specimens of early *Homo*.

The contribution from Louis Leakey (1903–1972) and Mary Leakey (1913–1996) to human evolutionary studies marks the beginning of the modern age of paleoanthropology because they immersed themselves in Africa as explorers and excavators, dedicated to the discovery of human origins. They found numerous sites. They inspired and literally started numerous careers of other scientists in search of human origins and evolution. They were excellent at publicizing their remarkable finds. They included local people in their expeditions and research. And most importantly, they solidified East Africa as the cradle of humankind.

Richard Leakey and Meave Leakey took over Louis and Mary's tenure in Kenya and continue to find exciting fossils today. Once it was established that Kenya and Tanzania are treasure troves of hominin sites, paleontologists began to discover a wealth of material in Ethiopia as well. 1974 marked the year of Donald Johanson's discovery of the *Australopithecus afarensis* skeleton called "Lucy" which was the most complete early hominin known to science at the time. Back in Kenya, during the 1970s, 1980s, and 1990s, a particularly successful group of paleontologists, nicknamed the "Hominid Gang," brought in a steady stream of hominin fossils. The team led by Richard Leakey (Louis Leakey's son), Alan Walker, Meave Leakey, John Harris, and Kamoya Kimeu collected numerous skulls and postcranial bones of *Homo erectus, Homo rudolfensis, Homo habilis, Paranthropus boisei, Paranthropus aethiopicus.* In 1985 they hit fossil pay dirt with the most complete early hominin skeleton ever found, a *H. erectus* boy called the "Nariokotome boy" or the "Turkana boy," named after the site and lakeshore where he was discovered.

Details of many of the most important hominin discoveries, especially more recent finds, are discussed in Chapter 3 where each hominin species is described.

Missing Links

There will always be missing links or gaps in the fossil record where we expect to find transitional forms. Organisms from fungus to fish to fishermen have evolved adaptations for recycling carcasses into energy, preventing the majority of organisms from fossilizing. The notion of "missing links" is perpetuated because each time a fossil discovery fills in a gap in the fossil record, two more gaps are created. Then once fossils are found to fill those gaps, four more are created, and so on (Figure 1.4).

Opponents of evolution use this seemingly endless quest for missing links to argue against evolution's credibility and paleontology's productivity. This misunderstanding is probably best termed "mything links" because there are, indeed, countless transitional forms on record from birds with teeth and bony tails (*Archaeopteryx*), to whales with little legs (*Ambulocetus*), to apes that walk upright (*Australopithecus*).

The terms "missing link" and "transitional form" are highly misleading as well. They imply that the organism was not well adapted and that it was merely waiting or even striving to eventually evolve into a modern form. They are also misinterpreted to mean that ancestral forms are somehow equal mixes or blends of their modern descendents.

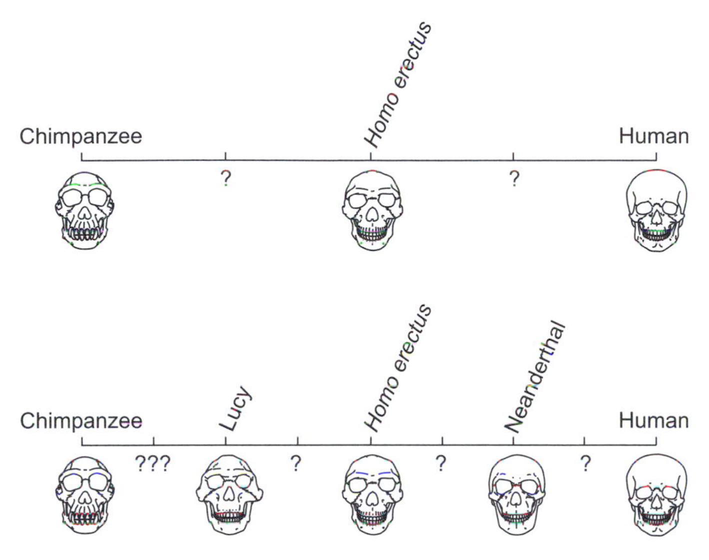

Figure 1.4 When Eugène Dubois found *H. erectus* he created room for two more missing links (above). One between *H. erectus* and humans and a second older, more primitive link joining *H. erectus* with the last common ancestor between humans and chimpanzees (LCA). Eighty years later, the *A. afarensis* skeleton called "Lucy" filled in the latter hole (below), getting one link closer to the chimp-human ancestor, "???", and to the other end of the chain at living chimpanzees. This idea of missing links can be misleading, however, because evolution is a branching process, not a blending or linear one. *H. erectus* is considered a missing link or a transitional form but it is not a blend of chimpanzees and humans, since it evolved after the split of the two lineages. *Illustration by Jeff Dixon.*

CURRENT ISSUES

Paleoanthropologists of the past and of today have discovered and continue to discover sites all over the Old World from Siberia to Australia, from England to South Africa, and from China to Portugal. Only relatively recently have scholars originating from Africa, Asia, Eurasia, Indonesia, and Australia joined the forefront of human origins science and it is certain that a brief history of human origins philosophy from their points of view would take different paths from the one rooted in Western Europe.

Table 1.1 Benchmark Discoveries in Human Origins and Evolution Science

1856	First *Dryopithecus*—Fossil ape from Europe
1856	First Neanderthal—Neander Valley, Germany
1868	First anatomically modern human fossil—Cro-Magnon, France
1891	First *Homo erectus*—Java, Indonesia
1924	First *Australopithecus*, "Taung Child"—Sterkfontein, South Africa
1959	"Zinj"—*Paranthropus boisei* skull, Olduvai Gorge, Tanzania
1960	Jane Goodall observes chimpanzees making and using tools in the wild
1961	First application of potassium-argon dating to hominin site—Olduvai Gorge
1967	Molecular clock theory is applied to human evolutionary studies
1974	"Lucy"—*Australopithecus afarensis* skeleton
1985	"Nariokotome boy"—*Homo erectus* skeleton
1995	First *Ardipithecus*—Oldest confirmed biped, Aramis, Ethiopia
1997	First ancient DNA of any fossil hominin (a Neanderthal) is analyzed
1999	Oldest bones butchered by stone tools are discovered at Bouri, Ethiopia
2000	*Orrorin*—Oldest purported biped, Tugen Hills, Kenya
2002	*Sahelanthropus*—Oldest hominin, Toros-Menalla, Chad
2000	Human genome is sequenced
2004	*Homo floresiensis*—The so-called "hobbits" of Indonesia
2005	Chimpanzee genome sequenced
2005	Only known fossil chimpanzee, Kenya

Many scientists currently working on human origins and evolution are only the first and second generation of intellectual offspring of the pioneers in the field. New fossil finds, new and improved dating methods, and new genetic analyses are constantly shaping and often overturning old ideas (Table 1.1).

Now that paleontologists have a considerable comparative fossil record of hominins, they are better at spotting the most fragmentary ones in the dirt or the mislabeled ones in museum drawers. And now instead of just a handful of fossil human ancestors, there are now literally thousands of specimens of at least eighteen different species (Figure 1.5).

Paleoanthropologists today are not as singularly focused on transitional ape-people as they were in the past. Since much of the hominin family tree has been filled in and since molecular (genetic) clocks have predicted the age of the split of the human and chimpanzee lineages, there is a priority on finding the earliest hominins and the earliest chimpanzees between 8 and 5 Mya.

With the expanding fossil record, the study of human origins and evolution can take place at many stages of prehistory and the field has broadened much further in scope than what Darwin, Huxley, and their cohort were pondering 150 years ago. The questions in Table 1.2 just

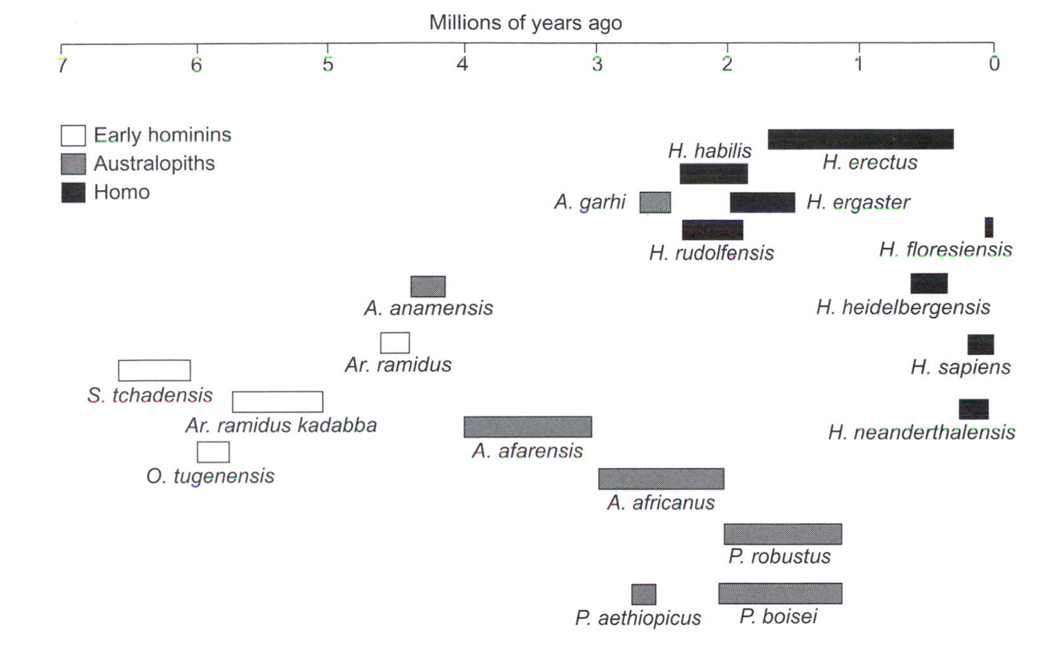

Figure 1.5 The hominin phylogeny. Lines linking the branches were not included because many of the relationships between the species remain unclear. However, there is good evidence that *Ardipithecus* (*Ar. ramidus kadabba* and *Ar. ramidus*) were ancestral to the australopiths *A. anamensis* and then *A. afarensis* in an evolutionary lineage. Prior to that scientists are still investigating *Sahelanthropus, Orrorin,* and *Ardipithecus* for evidence of their belonging to either the human or the chimpanzee lineage since they are located so close to the time when the two modern species are estimated (by molecular clocks) to have split about 6 Mya. *A. afarensis* is normally rooted at the base of both the *Paranthropus* and the *Homo* lineages, but sometimes *A. africanus* is put on the direct line to *Homo* too. The genus *Paranthropus* (*P. aethiopicus, P. robustus,* and *P. boisei*) is an evolutionary side branch from the direct lineage leading to humans that went extinct around 1 Mya. *H. habilis* and *H. rudolfensis* may have just been one highly variable species and *A. garhi* looks like it was the ancestor to these earliest members of the human genus. Some scientists lump *H. ergaster* fossils in with *H. erectus* but some keep it separate as it is shown here. *H. heidelbergensis* is an Archaic species that was probably involved in the evolution of Neanderthals (*H. neanderthalensis*) and modern humans (*H. sapiens*). *H. floresiensis* was a small-bodied and small-brained hominin known only so far from the Indonesian island of Flores that overlapped in time and space with modern humans, but may be dwarfed descendents of *H. erectus.* It will be helpful to refer to this figure as you read the descriptions of the species in Chapter 3. *Illustration by Jeff Dixon.*

Table 1.2 Current Questions in Human Origins and Evolution

- Is *Homo floresiensis* a valid species of dwarf hominin or is it diseased?
- Did modern humans around the Old World evolve in place from earlier hominin forms or did they evolve in Africa and spread to other regions, replacing those forms?
- Were Neanderthals directly involved in our own evolution? Did modern humans and Neanderthals interact?
- Why did the Neanderthals go extinct?
- Why did the robust australopiths—the genus *Paranthropus*—go extinct?
- Would we be able to recognize the earliest fossil hominin if we found it? How would we distinguish it from the earliest fossil chimpanzee?
- Why did bipedalism evolve?
- To what extent did meat-eating shape human evolution and enable brain size to increase?
- When did language evolve?
- Why are some traits, like darkly pigmented skin, more prevalent in some human populations compared to others?
- Is natural selection still affecting human populations? Are humans still evolving?

highlight a few of the popular research topics in the field. Fundamental concepts and evidence used to address these questions are introduced throughout the rest of the book.

2

From Fish to Fishermen

EVIDENCE FOR EVOLUTION

Everyone on Earth is related to everything that is living now and that has ever lived in the past. Fourteen billion years ago the Big Bang formed our universe and then, ten billion years later, matter came together to form planet Earth. Soon after that, the first proteins formed, possibly at thermal vents deep in the ocean. After that prokaryotes (single-celled organisms) and then eukaryotes (multicellular organisms) flourished. By the time we witness fossils of the "Cambrian Explosion," most notably from the Burgess Shale in British Columbia, there was a radiation of marine animals, many resembling modern crustaceans and insects. From those animals evolved the first animals with backbones, the chordates and the vertebrates. Lobe-finned fishes gave rise to the first tetrapods that managed to make a living on land by 385 Mya. Then ancient reptiles gave rise to dinosaurs, birds, and mammals. From some of the earliest mammals arose the group to which we belong, the primates. As Table 2.1 illustrates, natural selection slowly and unknowingly assembled humans by accumulating traits over millions of years of evolution. Charles Darwin grasped this notion well before there were many fossils to support it and well before any sort of plausible mode of inheritance of these traits was understood.

During his five-year voyage on the H.M.S. *Beagle* (1831–1836), Charles Darwin (1809–1882) collected evidence that he would use to construct his theory of natural selection. He gathered animals, plants, and fossils, and cultivated the ideas that he used to formulate a mechanism for evolution and natural selection. Looking back through a modern lens, the evidence Darwin observed is overwhelmingly strong today.

Table 2.1 Origins of Some Major Human Traits

Trait	Millions of Years Ago
Writing	0.006
Agriculture	0.01
Language	0.1
Symbols	0.1
Modern brain size	0.5
Meat-eating	2.5
Stone tools	2.6
Small canines	3.2
Bipedalism	5
Tricolor vision	23
Dental pattern	35
3 ear ossicles	130
Hair	150
5 digits	340
4 limbs	385
Jaws	460
3 semicircular canals	470
Backbone	520
2 eyes	550

Biogeography

Sailing around South America and eventually the world, Darwin witnessed species of animals and plants that very few Europeans had ever seen, many of which were unknown to European scientists. He noticed that not only did there seem to be regional similarities between organisms, but that organisms in neighboring environments tended to resemble one another rather than organisms in similar environments elsewhere in the world. In other words, organisms evolve locally.

A famous example of this biogeographic phenomenon, once noted by Darwin, is the variety of finches on the Galapagos Islands, an isolated archipelago created by volcanism and characterized by arid, rocky terrain. The finches on the Galapagos resemble the finches on the lush west coast of South America more than they resemble birds that exist in other harsh, arid environments.

Because islands lack whole groups of animals, newcomers have the opportunity to fill the open niches that are often already filled in their original environments. The original Galapagos finches competed with very few animals on these relatively isolated islands and were able to radiate into many species. They evolved a range of beak sizes and shapes, body sizes and shapes, and feeding behaviors. There are finches that behave and look like woodpeckers, but their genes link them to other finches of Galapagos, not to real woodpeckers elsewhere. Similarly in humans, it appears from genetic analyses that Melanesians with darkly pigmented skin are more closely related to Asians in that region of the world than they are to darkly pigmented Africans.

Fossils and Geology

Since the days of Lyell it is common knowledge that organisms go extinct and have gone extinct in the past and sometimes leave evidence of their previous existence in the form of fossils. The fossils that are deepest are the oldest and contain traits that link them as ancestors to later species. The primate fossil record is just one great example. The earliest primates are found in older rocks (65 Mya) than the first fossil monkeys, which are found in older rocks (35 Mya) than fossil apes (20 Mya). Only in the rocks that were formed within the last 6 million years do paleontologists find upright-walking ancestors of modern humans. Through time, as traits for bipedalism and other human characteristics like large brains increase, the ape-like ones decrease.

Artificial Selection

A fan of pigeon breeding, Darwin used evidence from the human domestication of animals to support his theory of natural selection. For thousands of years humans have selectively bred animals and plants, favoring the tastiest, most colorful, woolliest, smartest, heartiest, or fastest running. Darwin called animal husbandry and agriculture "artificial selection" because of how well humans can mimic natural selection. As a result of artificial selection, new varieties and new species are created. Dogs in all their splendid variety are the product of artificial selection from an ancestral wolf species.

Humans do not necessarily select for the same traits that natural selection would favor. Survival can be enhanced by human intervention, so traits that would potentially be detrimental to the success of an organism in the wild are able to thrive. For example, seedless watermelons are unable to reproduce and some fancy dog breeds need frequent medical attention for their cute, yet health-threatening traits like too short legs and too short snouts.

Homology and Analogy

The relationship between form and function in animals reveals how ancestry, or shared origin, affects the forms of animals. For instance, the human, whale, dog, horse, and bat forelimbs are similar despite their

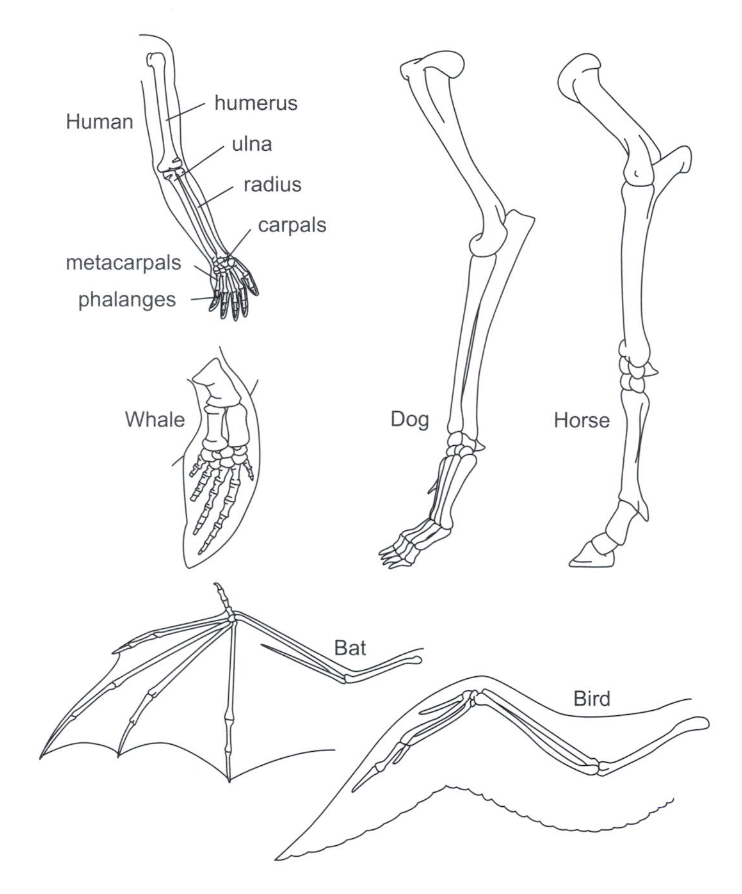

Figure 2.1 The human arm and the forelimbs of a whale, dog, horse, and bat are homologies because they all evolved from a common ancestral tetrapod. However, the wings of bats and birds are considered analogies because they did not evolve for flight from a common flying ancestor. Instead they evolved flight adaptations independently and their anatomy shows it: bat digits spread apart skin to catch lift and birds create the same effect without a bony infrastructure, just with their feathers. *Illustration by Jeff Dixon.*

different functions for manipulation, swimming, running, and flying (Figure 2.1). These structural similarities, or homologies, are due to their common, albeit distant mammalian ancestry and persist despite the evolution of different functions that change them through time.

Conversely, the wings of bats and birds represent analogies. Despite their similar function, the structures are not equivalent. Bat wings evolved from ordinary quadrupedal mammalian forelimbs and feathered bird wings evolved from bipedal dinosaur forelimbs. The similar functions evolved independently in what is called convergent evolution, or adaptive convergence, and they do not indicate recent relatedness or inheritance from a common ancestor. Bird legs and human legs are another example: both permit the animal to walk bipedally, but their foot anatomy is arranged completely differently. In many birds, the first toe (and sometimes the second one as well) points backwards for balance and grasping ability, but the big toe of a human sticks out in front with the others.

Animals with similar structures are more closely related despite seemingly vast differences in the modification and use of those structures (e.g., bats and humans) than many animals with similar functions are (e.g., bats and birds).

Vestigial Traits

Small, nonfunctional third nipples are not unusual in humans. These kinds of evolutionary leftovers, like the rare occurrence of a human tail, are what are known as vestigial traits, or atavisms. As the manifestation of an organism's evolutionary history, vestigial traits are stamps in the passport of evolution. Some mammals, like mice and dogs, have multiple mammary glands for feeding litters of young. Humans, like most primates (but there are some exceptional strepsirhines with multiple nipples), normally give birth to one baby at a time that requires only a pair of mammary glands for nursing. Extra nipples on humans symbolize the common ancestry with other mammals that have more than two teats.

Nipples on male mammals are not vestigial traits, as they do not signal an ancient state of male milk production. Male nipples are simply the result of an interesting developmental process. All humans begin development as a female and not until about the seventh week does an embryo with a Y chromosome start developing as a male. This transition occurs after the development of the skin on the chest, but before the development of the mammary glands underneath it. Male nipples are simply stamps in the mammalian passport of development.

Embryology

An understanding of the developmental processes of embryos is crucial to understanding evolution and the relationship between organisms. More closely related species have more similar developmental stages and developmental processes. Young human embryos resemble fish and actually have structures that look like gills. In fish these structures develop into gills, but in humans they develop into the ear, jaw, and neck.

Ernst Haeckel developed a hypothesis about vertebrate embryo development that was widely accepted by the eminent scientists of the early 20th century. It was called "ontogeny recapitulates phylogeny," which means that growth and development repeats evolutionary history. His hypothesis suggested that during development an organism experiences all the stages of its evolutionary history until it reaches full development. In this scenario humans first develop into fishes, then amphibians, then reptiles, then rodent-like mammals, then babies.

Haeckel's hypothesis is no longer accepted, but his general observations still ring true: The earliest stages of development *in utero* (i.e., in the womb) resemble primitive ancestry, and the later stages resemble more recent shared ancestry.

NATURAL SELECTION

Although we associate evolution with Charles Darwin, he was not the first to realize evolution occurs, nor was he the first to search for the mechanism. Jean Baptiste Lamarck (1744–1829) made the first robust, scientific attempts at explaining the Tree of Life. Much of Darwin's thinking was influenced by the writings of Lamarck and his ideas on the inheritance of "acquired traits."

The classic example of Lamarckian evolution is the explanation for the evolution of the long giraffe neck. Under this view giraffes had short necks originally and during their lives they stretched to reach leaves hanging high on the trees. Their necks became slightly longer than if they had not eaten those out-of-reach leaves. The offspring of the neck stretching proto-giraffes inherited these slightly longer necks and so on and so forth until after many generations the modern giraffe with a long neck was the result. Imagined in human terms, it is easier to see the flaw in Lamarckian evolution. A woman that lifts weights during her lifetime and builds up her muscles will not give birth to muscle-bound children, just as giraffes that stretched their necks will not produce longer-necked offspring. Environmental influences are not passed along to future generations, so unfortunately for Lamarck, without a grasp of genetics and inheritance in those days, his hypothesis fell short.

Although not the first to recognize evolution, Charles Darwin was the first to conceive of a feasible mechanism for it. Strictly speaking, Charles Darwin and Alfred Russel Wallace (1823–1913) independently came up with the theory of natural selection, but Darwin is given the majority of the credit because he published a widely read book. In *The Origin of Species* (1859), Darwin proposed a mechanism for evolution that—although it is the basis of modern evolutionary science—just cracked the surface for the discoveries that would follow in support of evolution.

Aside from evidence he gathered on the voyage of the H.M.S. Beagle, Darwin's ideas were also influenced by the work of economist Thomas Malthus (1766–1834). In his *An Essay on the Principle of Population* Malthus described how human population growth is exponential or geometric while food supplies can only grow at a relatively stable, linear, or arithmetic rate. Populations would grow continuously for eternity if limiting resources like food did not keep them in check. A fast growing population that reproduces too quickly for its resources can outstrip those resources and then experience a crash until the population size settles to a level the resources can support.

To preface his theory of natural selection, Darwin postulated the following: (1) The ability of a population to expand is infinite. But the ability of any environment to support a population is finite. He called this a "struggle for existence" and the "economy of nature"; (2) Organisms within populations vary and this variation affects an individual's ability to survive and reproduce; (3) Variation is transmitted from parent to offspring. Darwin called this "descent with modification" but at the time there was still no known mechanism of inheritance.

Put into modern terms, Darwin's theory of natural selection can be broken down into four parts.

Variation

Organisms vary. Each individual has its own phenotype, or suite of physical features like eye color, leg length, and vocal tone. Phenotypic traits include morphology (things like anatomy, size, shape, color, and texture), physiology, and behavior. Phenotypes are formed by both nature (genetics) and nurture (environmental influences like nutrition during development and activities during life).

Heredity

Some variation is inherited from the parents. Offspring resemble parents more than they resemble others in both physical appearance and behavior. The genotype, or genetic makeup, of an individual is linked to, and therefore influences, the phenotype. A mutation, or change in the DNA of an individual, can be beneficial or harmful and if it occurs in a gamete (egg or sperm) it can be passed on to the offspring.

Differential Fitness

Fitness in evolutionary terms is not physical fitness, like muscle tone and health. Fitness is the number of offspring an individual produces which is the equivalent of its reproductive success.

Some variations are more advantageous than others. Organisms with advantageous traits will survive and have more offspring who will carry those advantageous traits and pass them on to their offspring. Differential reproduction occurs when individuals with advantageous traits reproduce more than those without the traits. The increased survival and reproductive rates of an individual's offspring contribute to that individual's fitness being greater than another's. This phenomenon occurs because a trait is being "selected for" or because it is being "favored" by natural selection.

Adaptation

Environmental pressure, like what Malthus described, will determine which traits are more favorable than others. Traits associated with a better response to the pressure are adaptive and they become legitimate adaptations when they increase so much in frequency that a majority of the population has the trait. Adaptations take hold when successive generations of individuals with the adaptive trait have increased reproductive success or fitness.

For example, giraffes that have long necks may be the only ones that can reach and eat the leaves on the highest tree branches during times of drought. These long-necked individuals will have offspring that have long necks that are also successful during times of drought, and so on.

Adaptations can arise from unexpected places. For example, having a long neck for eating out-of-reach leaves could, in a new situation, be adaptive for something completely different like keeping the head above water in a flood. In this hypothetical case, having a long neck for feeding is an exaptation for swimming. It is hypothesized that the human larynx is an exaptation since it may have been co-opted for making speech sounds after it had already dropped in the throat for other reasons.

Certainly, the state of lacking a trait can be an adaptation as well. The loss of hind legs in whales led to their streamlined bodies for swimming. It is not yet known whether the loss of the ape and human tail offered an advantage or was simply the result of happenstance. After all, having a tail is beneficial for balancing on tree branches, and as a consequence of being tailless, apes do not move about the same way that monkeys with tails do.

THE HUMAN APPENDIX

In humans, the diminutive appendix is, presumably, not functional or necessary for passing food through the digestive system. It is smaller than it is in other mammals, but the human appendix has a tendency to rupture and if it is not removed promptly a person can die as a result. One may wonder, then, why the appendix has not disappeared completely. There may be a function that is yet unknown, like during development *in utero*. Perhaps, however, the appendix is not disadvantageous enough for selection to act strongly against it and remove it from the human phenotype. Few people die of a ruptured appendix. And many of those who do have already reproduced and passed along their genes that include blueprints for building an appendix in their offspring.

FROM MENDEL TO THE MODERN SYNTHESIS

Darwin was not able to fully realize his theory of natural selection because at the time no one understood the mechanism of inheritance, the force behind "descent with modification." Darwin laid down the foundation for evolution by hypothesizing that traits are passed on to offspring and if a trait is advantageous it will increase in frequency in a population because the offspring with the trait will survive and reproduce better and will pass that trait onto their offspring and so on. Accumulation of many traits like this will change a population from their original state to a different one in successive generations. However, no one could explain how those traits could be carried over in future generations.

Enter the Austrian monk Gregor Mendel (1822–1884). Using pea plants in breeding experiments, he worked out the missing piece, the unknown mode of inheritance, which allowed evolutionary theory to flourish and grow to where it stands today. Mendel remained unknown to evolutionary biologists for thirty-five years after he published his results.

Before genes were discovered, concepts of heredity included the medieval idea of "preformation" in which the gametes contained a homunculus, or tiny, complete human. Then, the notion of blending inheritance, which was supported by Lamarck and Darwin, among others, postulated that the heritable material from the parents is blended together in the offspring. Problematically, this idea contradicts the maintenance of variation that is clearly occurring in populations. It predicts that traits would be muddied or diluted out. For example, a dark-skinned mother and a light-skinned father would always blend to have medium-pigmented children and they, in turn would only offer medium pigmentation to their children, eventually leading to a uniform skin color in future generations. In reality, those parents can produce a spectrum of pigmentation in their children and they in turn can produce children that look like themselves, like their mate, like their own parents, or like their mate's parents.

These and other early ideas, like Darwin's pangenesis, were dropped once Mendel's experiments were brought to light. Although what Mendel discovered is an oversimplified view, heredity is perhaps easiest to understand through his work. Mendel's experiments were scientific. He kept track of the numbers of offspring and their phenotypes which resulted from different plants bred together. He also performed a large number of crosses to lessen the effects of chance and this enabled him to replicate experiments as well. He chose to observe contrasting, binary traits (e.g., yellow/green and smooth/wrinkled), which are easy to track through generations, because they are discrete traits, as opposed to continuous ones like height or human skin color. According to the rules of good science his hypotheses were falsifiable and then useful for predicting further observations. For instance he found that when he bred true strains of yellow and green pea plants the ratio of offspring was nearly always three yellow to one green. These types of experiments helped him realize a few simple rules of simple inheritance put here into modern terms based on current understanding (e.g., the term "gene" was not introduced until 1909).

1. Two gene variants, or alleles, one from each parent, determine the phenotype. The phenotype will be produced by the dominant allele, not the recessive one. Alleles are not blended. They are inherited and expressed separately.
2. Each parent has two alleles for each trait, and the chances are equally likely that offspring will receive either allele from the parent. This notion is what is known as Mendel's "Law of Segregation." At conception, when the zygote is made from the union of sperm and egg, alleles from the mother and the father segregate randomly into daughter cells so there is a predictable 50 percent chance of getting one allele or the other. This process contributes to the maintenance of variation in a population for natural selection to act upon, since no two offspring will have the same combination of alleles.
3. Traits are inherited independently. That is, pea color was not affected by whether or not the pea was smooth or wrinkled. This is known as

Mendel's "Law of Independent Assortment" and is another means of perpetuating variation in populations.

Although Mendel's explanations of inheritance (based on carefully chosen traits) proved to be oversimplified, Mendel was a pioneer who laid the foundation for the whole of modern genetics. It is widely and wildly apparent that Mendelian-style inheritance is the exception rather than the rule. Today it is well known that the majority of traits are complex and expressed by multiple genes that are also often linked in inheritance. A deep understanding of modern developmental, cell, and molecular biology is required to comprehend the complex modes of inheritance of most traits.

Mendel published his findings in 1866, just six years after Darwin's treatise. It went unnoticed by the scientific community until 1900, when finally, a mechanism of inheritance could be married to natural selection and shortly thereafter a theoretical revolution known as the "Modern Synthesis" was born. Then, once James Watson, Francis Crick, and Rosalind Franklin rendered the first accurate model of the DNA molecule in 1953, the fundamental genetic component of evolution was revealed.

DNA, CHROMOSOMES, CELLS, AND INHERITANCE

> The basic plan for the cell contained in the genome . . . work[s] so well that each human develops with few defects from a single fertilized egg into a complicated ensemble of trillions of specialized cells that function harmoniously for decades in an ever-changing environment.
>
> —Pollard & Earnshaw, 2002

Individuals, or organisms, are built of organs, which are made of tissues, which are comprised of cells. Humans (and everything but bacteria and Archaea) are made of eukaryotic cells that contain nuclei (sing. nucleus). All types of cells fit into two main categories: somatic cells and gametes. Cells are separated into these two groups according to their different functions, their different modes of cell replication, and their different amounts of genetic material in the nucleus.

The nucleus of a cell contains the genetic material in the form of chromosomes, which are very condensed, long strands of DNA (deoxyribonucleic acid). Each nucleus in a human cell contains 23 pairs of chromosomes. There are 22 pairs of autosomal chromosomes

(44 chromosomes) and one pair of sex chromosomes (2 chromosomes) for a total of 23 pairs (46 chromosomes). Females have two X sex chromosomes (XX) and males have one X and one Y (XY). Chimpanzees have 24 pairs of chromosomes and humans have one pair less because two of them fused during evolution.

Gametes contain exactly half the complement of chromosomes in somatic cells. That is, they contain one of each of the 23 chromosomes, or 23 unpaired chromosomes. During fertilization, when the sperm and the egg unite and form a zygote, each gamete contributes half the necessary DNA to produces cells with the complete set of all the pairs of chromosomes, enabling it to develop normally into an embryo and beyond. When the offspring develops further it will eventually produce its own gametes that contain half of its genetic material to be passed on when it reproduces. Then the cycle of life continues.

Somatic cells need to divide and duplicate to grow tissues during development and to maintain the tissues when old or damaged cells die. Somatic cell division, or mitosis, is constantly occurring throughout the human body at the estimated rate of 300 million cell duplications per day, which slows down with age. Mitosis results in two copies of the parent cell.

Gametes divide to make the creation of offspring possible. Meiosis, the process of gamete division that occurs in the ovaries in females ("oogenesis") and the testes in males ("spermatogenesis"), results in four daughter cells. Each daughter cell inherits only half the chromosomes of the parent cell. For female humans, meiosis occurs in the developing fetus and by the time they are born little girls possess all the eggs they will have for the rest of their lives. Contrarily in males, meiosis continuously replenishes sperm throughout most of the life span. All eggs carry the X chromosome (since the female parent cell has two Xs) but sperm can carry either the Y or the X. It is therefore the male contribution to the zygote that determines the sex of the offspring, since if a sperm carrying a Y chromosome fertilizes the egg, the baby will be male (XY) and if it is an X chromosome the baby will be female (XX).

Crossing-over of the chromosomes is a phenomenon that only occurs during meiosis. Parental DNA gets swapped and the resulting chromosomes contain a mosaic of genes from the mother and the father. As a consequence, a chromosome can have one paternal end and one maternal end. The offspring has new combinations, or a recombination, of genes from both the mother and the father that comprise a brand new combination that does not exist in either of the parents. Crossing-over and recombination are crucial for maintaining genetic variability in successive generations because offspring are far from being clones of their parents.

DNA is commonly called the "genetic code" or the "genetic sequence." The DNA molecule is comprised of two strands of DNA sequences twisted together like a spiral staircase (called a "double helix") with steps made of nucleotides. For simplicity, these nucleotides are symbolized by letters that "spell out" the DNA sequence: A (Adenine), C (Cytosine), T (Thymine), and G (Guanine). These nucleotides or "base pairs" bond and hold together the double-stranded DNA molecule. Adenine only binds to thymine (A-T) and cytosine only binds to guanine (C-G). For example, the sequence of ATAG binds to TATC in the other strand.

Genes are segments of DNA, or pieces of the code, that are spelled out by these letters. Genes code for amino acids that build proteins. Proteins make up everything in the body, including structures, fluids, hormones, and even enzymes that catalyze, or enable, reactions to build other proteins. The chromosomal loci (sing. locus) of many genes are known and when one person has a different arrangement of nucleotides for that gene than another person they are said to have different alleles, or variants, of a gene. For instance, there are two alleles for type of earwax in humans, one dry and one wet, but there is only one gene for earwax (*ABCC11*). Some genes have more than two alleles in a population, like the gene for blood type which has alleles A, B, and O.

An individual can only carry two alleles for any given gene; the one on the chromosome they inherited from their mother and the one on the chromosome they inherited from their father. In many cases, but not all, one allele will be dominant (A) with respect to the other that is recessive (a). The dominant allele (A) will be expressed in homozygous, (AA) or heterozygous (Aa) individuals where it masks the expression of the recessive allele. The recessive allele will be expressed only in homozygous individuals for that allele (aa) because there is no dominant allele to mask it. In Mendel's experiments, the allele for yellow was dominant and was expressed in both "AA" and "Aa" individuals and the allele for green was recessive so it was expressed less frequently because it could only occur in "aa" plants. Beyond Mendel we know that alleles can be codominant (e.g., Sickle Cell Trait, Chapter 4) and they can also vary in expression on an individual basis depending on the presence of other linked alleles or because of alleles that regulate their expression levels.

For genes to be properly expressed, DNA must get read and copied properly in cell division and must get read and translated properly during protein construction. If any nucleotide is read incorrectly in the sequence, the wrong amino acid can be called for, which could build a totally different protein. Fortunately there are checks and balances built into the system to prevent such reading errors from causing disaster. For instance, for any single amino acid, there are up to four different (albeit very similar) nucleotide arrangements that call for it. This redundancy in the code, or error tolerance, is how many reading and translating errors (i.e., mutations) are prevented from being harmful or lethal. Code redundancy is also how mutations in the sequence can remain neutral and cause no harm to an individual, since a new arrangement of nucleotides does not always express a different protein.

FORCES OF EVOLUTION

Natural selection acts on the individual but evolution occurs at the population level. Changes in allele frequencies in a population over time are referred to as microevolution, which can occur quickly and is directly observable. On a grander scale, macroevolution is long-term change, characterized by speciation that mostly occurs over deep, or geologic, time. Accumulated microevolutionary change results in macroevolutionary change. Microevolution and macroevolution lie on a continuum of change from small- to large-scale, both resulting from the same genetic mechanisms and the same four forces: mutation, gene flow, genetic drift, and selection.

Mutation

Random mutation provides new material for natural selection. Changes in the genetic code are passed on to offspring when random reading and copying errors occur during meiosis. Mutations also occur through exposure to sequence-changing, mutation-inducing agents ("mutagens") like radiation and chemicals. Most mutations cause spontaneous abortions, are harmful or are simply neutral. Beneficial mutations are rare, but when they occur in the gametes and can be passed on to offspring they can make a large evolutionary impact.

Point mutations, which are nucleotide deletions, substitutions, or additions, are the most common kind of change in genetic code. Mutations can involve whole or partial chromosomes as well. These errors occur during meiosis when eggs or sperms receive too many or too few chromosomes and then get fertilized. The result is often miscarriage. Trisomy 21, or Down Syndrome, is one of the few conditions that is viable. It occurs in one out of a thousand live births. Individuals have three instead of two copies of Chromosome 21 and develop a unique, recognizable face and head shape, are normally smaller in stature, and have varying degrees of developmental and health problems.

Gene Flow

Alleles are exchanged between populations through gene flow which can occur between neighboring populations or when populations are new to a region because of dispersal or deliberate migration. Human gene flow is regulated by culture, which determines how frequently populations interact and interbreed and also how they move around through space and time. Gene flow mixes the alleles of different gene pools, thus preventing them from diverging into separate species.

Genetic Drift

Random change in allele frequency through time is genetic drift, which is a state marked by the absence of gene flow. Drift is caused by chance alone and the result of very strong drift is either allele extinction or fixation. That is, alleles can be completely eliminated from the gene pool, while others can take hold in each and every individual. Drift is most commonly caused by a population bottleneck or by a founder effect and is most observable in small populations.

A population experiences a bottleneck when its size decreases relatively rapidly and then increases again, as the result, for instance, of famine, war, genocide, or disease. A founder effect is a type of bottleneck that occurs when a small subgroup is isolated from the population and begins a new population by mating with one another, like if, for instance, a small group of tourists got shipwrecked on an uncharted desert island. With drift, neutral traits like six fingers (polydactyly) or maladaptive traits like Ellis-van Creveld Syndrome (a disorder that involves limb dwarfism, polydactyly, and heart and bone malformation and exists at higher than normal frequencies in the Old Order Amish of Pennsylvania) are perpetuated simply because of chance and small population sizes. Drift results in a decrease in the variation within a population or a gene pool.

Selection

As the strongest evolutionary force, selection has the ability to retain or eliminate alleles in a population. Favorable alleles spread. Unfavorable alleles disappear.

For example, a new mutation or a neutral trait that already exists becomes advantageous under environmental conditions and it enables individuals to be more successful than those without it. Humans are generalists and can thrive in a wide variety of conditions, but all organisms are influenced by changing environmental conditions, some are just more vulnerable or sensitive than others.

Selection favors individuals with the "best" traits, but if conditions change, traits that are neutral or even adverse one day could be advantageous the next. Because of the variation maintained by meiosis, recessive traits, mutation, and gene flow and because of the chance retention of traits due to genetic drift, neutral and undesirable traits will always persist in populations. If conditions change dramatically or quickly, this variation will save a species because natural selection will have more to work with before that species is lost to extinction.

Hox Genes

Our bodies are made up of patterned segments controlled by correspondingly patterned and segmented genes, or Hox genes. Changes in the number of these repeated parts (through mutations that result in gene duplication, gain of gene function, gene deletion, or loss of gene function) can perhaps explain evolutionary "jumps" like the gain and loss of legs in various lineages (whales, snakes, legless lizards), and the addition or loss of teeth, fingers, and toes. Mutations affecting the regulation of Hox genes will also cause developmental innovations and these include changes in timing, location, and quantity of Hox gene expression.

Evolution of the vertebral column is of considerable importance to human origins. The segments of the backbone (vertebrae) have undergone changes over the course of primate evolution, most of which are attributed to changes in locomotion. Apes have six lumbar vertebrae (in the lower back) but humans normally have five. Also, both apes and humans have lost the tail all together (the very end of the vertebral column). Mutations in Hox genes are probably responsible for these changes.

SEXUAL SELECTION

Darwin conceded that natural selection could not account for all evolutionary change. Even if modern concepts of mutation, gene flow, and genetic drift are considered, they cannot readily explain traits like the peacock's cumbersome tail, which baffled Darwin into quipping that the mere sight of it nauseated him.

To account for puzzling sorts of excessive and potentially handicapping traits, Darwin conceived of another type of selection, sexual selection, where the males and/or females of a population select mating partners based on characteristics which do not always have a conspicuous link to fitness.

Sexual selection is shaped by the difference between males and females in reproductive rates. Male mammals have a higher rate than females; that is, they can produce more offspring per unit of time. Females must wait to conceive again for weeks or months while carrying a fetus to term, but moments after conception males can conceive again with a different female.

There are also differences between sexes within species in parental investment strategies. In some species the male invests little in the offspring beyond his initial contribution of sperm, while the female is the sole caretaker from conception until the offspring is mature. Because the female carries the fetus and produces the milk, the female strategy is to invest largely in time, health, nutrition, and energy. Because the male does not carry the fetus or produce milk, the male strategy is best if he mates with numerous partners. The male strategy is further adaptive since male paternity is unknown without genetic testing. Males have alternative strategies in species that form monogamous pair-bonds, like gibbons, where males contribute largely to the offspring through territory protection and mate guarding.

In systems with sexes that have differing strategies, females tend to be the choosy sex. They make the most parental investment, take the most risk, and are the limiting resource for male reproductive success. Therefore, males are the competitive sex. They compete with other males for access to females. From an evolutionary standpoint, this scenario favors phenotypes that give a mating advantage to males, even if they lower their chances of survival. The most successful males are, for instance, the most aggressive, the strongest, the most beautiful, or the longest in canine teeth (competitive weaponry). The most successful females in such a system are the best choosers of males. Part of Darwin's difficulty in conceiving of sexual selection was cultural; he understood male competition but had a weaker grasp of female choice.

Sexual dimorphism—the differences in the size, color, or anatomy between males and females of a species—is exaggerated in species where mating strategy differences are exaggerated and mate competition is high. In monogamous apes like gibbons, the males and females are very hard to differentiate without close inspection. Their canine sizes and body sizes are the same and males have small testes because the need to produce copious amounts of sperm is diminished in the absence of competition with males.

Chimpanzees, however, have much larger testes because they compete for females in a multi-male, multi-female mating system. Chimpanzee males and females show differences in body size and canine size, with males having larger traits over all. In the gorilla system, males and females show even more body size dimorphism than chimpanzees. The dominant gorilla males are known as "silverbacks" because they display silver shocks of hair down their backs. Silverbacks usually have sole access to a group of females. Conversely, testes size in gorillas is very small which may correlate to success in preventing other males from mating with the females in the group.

Females choose mates that offer direct benefits to her like protection, food, and parental care. They can also gauge a male's fitness indicators which signal whether or not he has good genes that will increase the fitness of her offspring. Such indicators in humans have been shown to be as simple as clear eyes, shiny hair, glowing skin, a symmetrical face and body, dancing ability, and so on, but fitness indicators can also seem frivolous, silly, or even burdensome like the peacock's tail. The evolution of such traits are explained by Zahavi's handicap principle which states that traits can evolve that have no apparent function other than to advertise the fitness of the individual. That is, some traits simply provide means for bragging that the individual is so strong or so healthy that it can overcome a handicap.

Fitness indicators, whether they are handicaps or not, evolve through a process called runaway selection, that was conceived by R.A. Fisher in 1930. Say, for instance that a female has a preference for a larger than average nose. She will mate with large-nosed males and produce large-nosed offspring. Her sons will raise the average nose size in the population and her daughters will carry her gene for preferring large noses. If the cycle is allowed to continue and if large noses are indeed correlated to fitness, females will select for larger and larger noses, thus increasing the average nose size of the population. Runaway selection accounts for the evolution of sexually dimorphic traits as well as the preference for them.

SPECIATION

Accumulated microevolution, or changes in allele frequencies in a population, leads to macroevolution, which is speciation. In this sense, variation that exists within a population eventually increases to become variation between populations. Evolution at the species level is the result of cumulative microevolution. Fishes did not and do not evolve into humans, instead a fish-like ancestor gave rise to all amphibians, reptiles, birds, and mammals, including humans. Step-wise changes accumulated through time in the fins of fish to become weight-bearing legs for walking, and the rest is (pre)history.

Allopatric speciation occurs when a geographic barrier like a river, a highway, or islands separated by rising sea levels isolates a subset of the population and the forces of evolution go to work independently on the two populations. The accumulated changes lead to reproductive incompatibility.

Parapatric speciation, a muddy model of speciation, occurs when a subset of a population that covers a large geographic distribution, over a variety of habitats, becomes isolated by distance and selection acts differently on the gene pools according to their different habitats. These populations can be held together as one highly variable species by hybrid zones where interbreeding occurs. Baboons, which are monkeys that range from North to South Africa, are a good example of this phenomenon as some scientists lump them into one diverse species, but others see the subpopulations as their own individual species.

Sympatric speciation is considered a rare mode of speciation because it occurs without geographic or physical boundaries. Sympatric speciation is the formation of two species living in the same place. Behavioral boundaries, like differences in vocalizations or courting rituals, that prevent full gene pool-wide interbreeding or gene flow can permit natural selection and sexual selection to act differentially within the population and result in a subpopulation actually becoming a separate species.

The types of reproductively isolating factors that can drive speciation include failure to recognize mates, behavioral differences, habitat preferences, timing, morphological or mechanical incompatibility during mating, nonfertilization of egg, and nonviable hybrid offspring. None of these things prevent gene flow between human populations despite the diverse and variable nature of the species.

Species are ultimately arbitrary classification categories placed on continuous, overlapping portions of the Tree of Life. There are many different definitions of species, or species concepts, which work for different organisms in space and time. The morphological and the genetic species concepts group organisms together that possess similar anatomy and genetic codes, respectively. The ecological species concept groups those that may have the same morphology but have a different niche. The biological species concept groups organisms together that can successfully reproduce viable, fertile offspring.

The best concept to apply to fossil organisms is debated. As of now there are no fossils of hominins killed in the act of mating. (Although there are dinosaurs from Mongolia that fossilized in the act of fighting and there is a fossil ichthyosaur (an extinct dolphin-like reptile) that got buried while giving birth.) In fossils too old to preserve ancient DNA it is impossible to know which ones could interbreed successfully. Therefore, the biological species concept, which is the concept most commonly used to group living mammals, is of no use for hominin paleontology. Instead, scientists must look to morphological variation within living species of humans, monkeys, and apes to gauge and predict the amount of variation one expects to find within species in the hominin fossil record.

Intelligent Design

On December 20, 2005, Judge John E. Jones ruled in the *Kitzmiller vs. Dover* trial that the Dover Board of Education in Pennsylvania could not mandate the teaching of intelligent design in science classes because it is not a science. Supernatural explanations are by definition outside the realm of science. However, supporters of intelligent design continue to lobby for its inclusion in the science curriculum of schools around the world.

Intelligent Design (ID) is an idea that posits that the universe is too complex to have evolved by natural selection and must have been designed by an intelligent entity. ID appeals to the spirit of American democracy by falsely claiming that there is a debate within the scientific community about evolution and that students deserve to hear both sides of the debate. But, this is misleading since scientific debates focus on details about the nature of evolution not about whether or not it occurs. Plus, science is not a democracy. All ideas are not given equal weight. We keep and build upon the ideas that have accumulated good scientific evidence and we discard those that are shown to be invalid. Science and technology would never progress if they did not operate this way.

Examples of unintelligent design are abundant, even in humans who are often considered the pinnacle of design. The human eye has blind spots and it sees everything upside down so the brain must flip everything right side up. Due to anatomical changes in the throat that are associated with speech, it is easy to get food lodged in the trachea (the "windpipe") and choke. The combination of the narrow birth canal (a pelvic adaptation for walking upright) and the very large head of brainy human babies makes childbirth difficult and often dangerous. Because of the constraints on the raw material that natural selection had to work with, some human traits appear to be jerry-rigged. But, we overcome them with other anatomical adaptations and with behavioral adaptations and even with cultural innovations like the Heimlich maneuver.

CLARIFYING EVOLUTION

It is difficult to comprehend deep time. Our narrow 100-year glimpse of the universe makes it difficult to imagine thousands, let alone hundreds of thousands, millions, or billions of years. Our 100 years at best is a mere 0.000000025 percent of the Earth's history. Alternatively, if the Earth's history is condensed into one year, humans arrive at the New Year's Eve party with less than thirty minutes to spare. Under these constraints it is difficult to witness large-scale evolutionary changes, like those that occurred between fish and fisherman.

Evolution may have no goal, but it is not random either. Selection acts on the available traits under the current conditions, has no memory of the past, and is not propelled by an inherent need to improve a species. Chance mutation, which introduces genetic variation, may be random but selection does not act randomly. Selection favors advantageous traits, not random traits. Selection can only work with what is already there and with what is developmentally feasible. It has no foresight to plan for inevitable changes in the environment that will change what it favors in future individuals under those new conditions. Through the four forces of evolution (mutation, drift, gene flow, and selection), well-designed traits are produced through millennia of trial and error.

Populations evolve, not individuals. Although the term evolution is often poetically applied to a person's physical or intellectual development throughout life, an individual cannot evolve. A six-toed baby born to five-toed parents is not the result of evolution unless six-toed people increase in frequency in that population over time in successive generations. Scenarios for supporting the increase of six-toeness in a population could stem from a founder effect, from a population bottleneck, or from selection preferring six toes to five.

Evolution is not the "survival of the fittest." Herbert Spencer (1820–1903) is credited for marrying this catchphrase with Darwin's theory of natural selection, but it is dreadfully misleading. As a consequence, natural selection is often misinterpreted as "only the strong survive" and, sadly, "the strong kill the weak." Fitness does not necessarily have anything to do with strength and its attainment does not necessarily require physical violence. Fitness is simply reproductive success. Individuals with favorable adaptations will survive and reproduce offspring with that favorable adaptation, thus increasing their fitness.

If there must be a catchphrase for evolution perhaps "survival of the *fitter*" is more appropriate. Evolutionary success is not measured in being the best. Being slightly more favored than others is enough. For example, on a continuum of short to tall people, those on the lower end of the spectrum, not necessarily the very shortest, are favored by selection if a general level of shortness is advantageous. In the next generation, the genes of the fitter ones increase in frequency compared to the less fit. The shorter individuals do not violently eliminate the tall ones from the population. The short ones simply out-survive and out-reproduce the tall ones. (This can be referred to as "out-competing," but this phrase also tends to paint an unfair kill-or-be-killed portrait of evolution.) If selection continues to act strongly on the shorter ones, eventually the tall ones will disappear or, if isolated and then favored by selection, the tall ones will continue to evolve as a separate lineage.

COLLOQUIALISMS

Evolution often gets accused of being "just a theory" because of the confusion between casual and scientific uses of theory. Everyday theories are hunches and are often muttered with tongue-in-cheek to mock the likes of Sherlock Holmes. On the contrary, a scientific theory is no harebrained scheme. It is a logically self-consistent model or framework for describing the behavior of a certain natural phenomenon supported by strong experimental evidence. It is a systematic and formalized expression of all previous observations made. It is predictive, testable, and has not been falsified. Although, just like any idea in science, it is open to falsification. When bolstered with enough evidence, scientific theories, like evolution, are considered facts.

The standard expression that one "believes in evolution" is an unfortunate consequence of the shortcomings of the English language. It sets up an antagonistic dichotomy between evolution and religion, implying that evolution is something to believe in as opposed to believing in a deity. No one says that they believe in gravity or electricity, and evolution should be treated accordingly.

TAXONOMY AND CLASSIFICATION

Categorization is a skill every human uses, even for matters that extend beyond zoological nomenclature. A formal system of classification of organisms, or taxonomy, is essential because it provides a language so that people can collaborate, understand one another's results, and test one another's hypotheses.

Not all classification schemes for animals translate across cultures or stand the test of time. For example, the classification of animals in an ancient Chinese encyclopedia includes the following groups: (a) those that belong to the Emperor, (b) embalmed ones, (c) those that are trained, (d) suckling pigs, (e) mermaids, (f) fabulous ones, (g) stray dogs, (h) those that are included in this classification, (i) those that tremble as if they were mad, (j) innumerable ones, (k) those drawn with a very fine camel's hair brush, (l) others, (m) those that have just broken a flower vase, (n) those that resemble flies from a distance (*Celestial Emporium of Benevolent Knowledge*, referred to by Borges, 1964).

Because of its logical and scientific approach, the classification system created by Carolus Linnaeus translates universally (Table 2.2). In the Linnaean system, the genus and species are always underlined or italicized and the genus can be abbreviated so that *Homo sapiens* becomes *H. sapiens.*

At the time Linnaeus named the class that includes humans, he was publicly lobbying for the benefits of breastfeeding, so he chose "Mammalia" to honor mammary glands. His alternative name was the equally appropriate "Pilosa" for "hairy things." So Linnaeus chose to name hairy animals based on a trait that is only functional in half the species and, in them, only functional during a fraction of life. However, what Linnaeus did not know at the time is that mammary glands actually evolved from the same types of follicles in the skin that hairs evolved from, so in the grand scheme of things none of this name game actually mattered.

The taxonomy of organisms implies relationships that can be used to build phylogenies, or trees of relatedness, with the groups in a nested hierarchy. Two species, or branches on a phylogeny, are more closely related to each other than a third, three species are more closely related to each other than a fourth, and so on. Species are considered closely related if there is evidence for their shared ancestry in either their phenotypes or genotypes.

Evidence for evolutionary relatedness comes in the form of shared, derived features that have evolved since the last common ancestor of the species under consideration. Lack of a tail is a shared, derived feature linking humans and apes to the exclusion of more distantly related monkeys. Primitive, or ancestral, features like five fingers and toes are not significant and contain no information as to the relatedness of

Table 2.2 Classification of Humans

Kingdom	Animalia	Animals
	Eumetazoa	Multicellular animals
	Bilateria	Symmetrical animals with right and left sides
Phylum	Chordata	Notochords/spinal columns
Subphylum	Vertebrata	Backbones
	Craniata	Having a skull: fishes, amphibians, reptiles, birds, mammals
Superclass	Gnathostomata	Jawed vertebrates
Class	Sarcopterygii	Lobe-finned fishes and terrestrial vertebrates
	Tetrapoda	Four limbs
	Amniota	Develop in an amniotic sac: reptiles, birds, and mammals
Class	Mammalia	Hair and mammary glands
Infraclass	Eutheria	Placental mammals
Order	Primates	Strepsirhines and haplorhines
Suborder	Haplorhini	Tarsiers, New and Old World monkeys, apes and humans
Infraorder	Catarrhini	Old World monkeys, apes, and humans
Superfamily	Hominoidea	Lesser apes, great apes, and humans
Family	Hominidae	Great apes and humans
Subfamily	Homininae	Chimpanzees and humans
Tribe	Hominini	Humans and their extinct bipedal ancestors, "hominins"
Genus	*Homo*	"Person"
Species	*Homo sapiens*	"Wise person"

humans, apes, and monkeys because all groups have five fingers and toes.

Convergent traits evolve in parallel in different lineages from a common ancestor that did not originally have the trait. For example, flight evolved in dinosaur descendents (birds) as well as in small mammals (bats).

Despite the complications that parallel evolution can bring to evolutionary phylogenies, parsimony is always the rule. The principle of parsimony is best summed up by the principle of "Occam's razor" which states that the simplest explanation is usually the correct one. With parsimony, phylogenetic trees are based on the fewest evolutionary changes and the fewest convergences.

Classification is ultimately arbitrary. A major problem with imposing taxonomy onto nature is that nature is not ordered into neat little categories. As one approaches the roots of lineages, it becomes more and more difficult to clearly place fossil species on either of the two or more branches that split around the time of its existence. For instance, the closer you get to the last common ancestor of humans and chimpanzees (LCA) in the fossil record, the harder it is to place fossils with confidence on either the chimpanzee or the human branch.

PRIMATES

With clothing, language, automobiles, and mobile phones, it is easy to forget that humans are primates (Figure 2.2). Anatomy and genetics indicate that primates as a group are most closely related to even less-humanlike animals: colugos (*Dermoptera*) which are gliding mammals from Southeast Asia nicknamed "flying lemurs" and tree shrews (*Scandentia*) which are tiny shrew-like animals that also live in Southeast Asia.

All primates, with a few exceptions, share general trends in behavior, brain size, single offspring (not litters), extended stages of growth and development (i.e., prolonged "life histories"), sociality, anatomy, grasping hands and feet and useful thumbs, nails instead of claws, forward facing eyes, stereoscopic vision, a generalized (i.e., versatile) body plan, generalized teeth, a variable diet, a bony case for the three ear ossicles ("auditory bulla"), and an enclosed eye socket ("orbit").

Primates are classified into two major groups, or suborders: Strepsirhines ("wet-nosed") include the lemurs and lorises, and Haplorhines ("dry-nosed") include the tarsiers, monkeys, apes, and humans.

Lemurs are the cat-like leapers ("vertical clingers and leapers" to be exact) that live only on Madagascar, the large island off the southeast

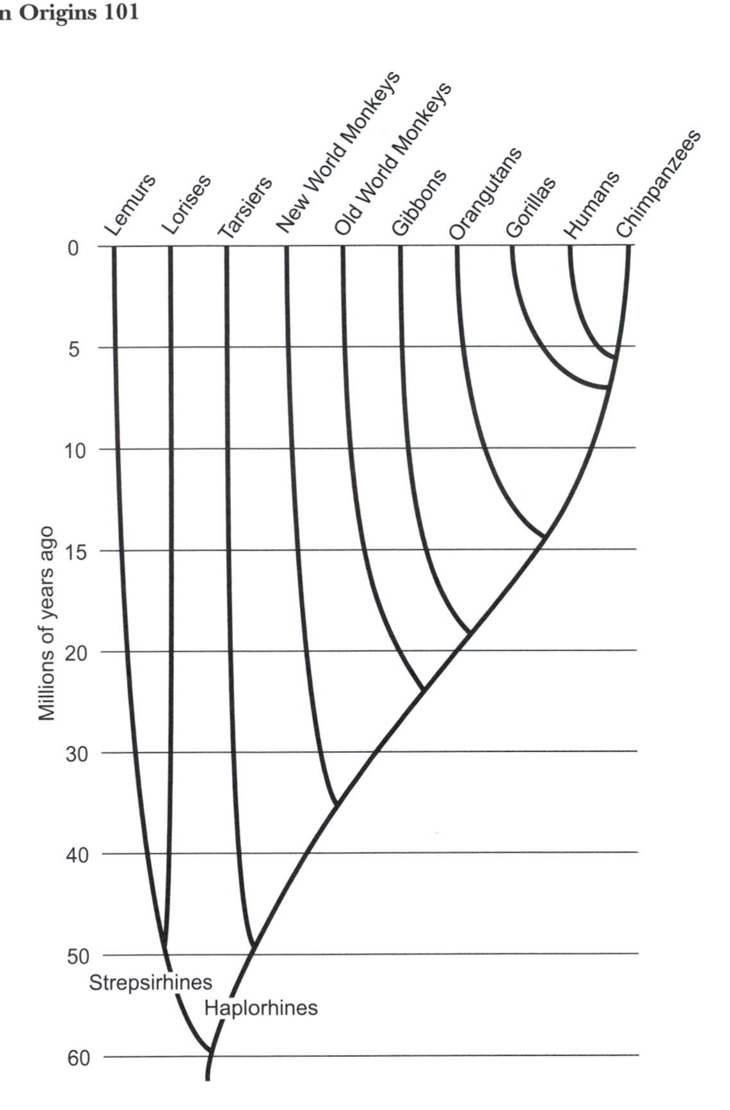

Figure 2.2 The primate Tree of Life shows approximate splitting times of the major groups. *Illustration by Jeff Dixon.*

coast of Africa. Lorises are a group on the African mainland that contains slow-moving pottos and leaping galagos or "bushbabies."

Tarsiers are somewhat difficult to group decisively with the lemurs and lorises or with the monkeys and apes. They have shared derived features with monkeys and apes in many different physiological and morphological complexes like in the eye (i.e., a light-sensitive pit in the retina of the eye for sharp vision ("retinal fovea") and the lack of the reflective back of the eye which enhances night vision, also known as the "tapetum lucidum" which is same thing as the shiny reflection of a dog's eye in flash photography). However, despite their derived features, tarsiers fit into the same level of adaptation as the lemurs and lorises and are sometimes grouped together into prosimians—a name that highlights the primitive, not-quite-monkey ways of lemurs, lorises, and tarsiers. Tarsiers have primitive features like an unfused jaw at the middle and grooming claws on some digits (instead of nails).

This problem with confidently grouping the two main suborders of primates boils down to the fundamental problem with classification. Forcing an organism into only one group when it shares features with two is always going to cause a problem simply because of the branching and overlapping nature of evolution. Organisms that evolve near the split of major groups will resemble both groups.

Primates are intelligent and have large brains relative to their body sizes. The major features of primate behavior—including group living, geographic boundaries, home ranges, prolonged mother–infant bond, long learning period, affiliations and coalitions (enhanced by grooming), and communication (vocal and nonvocal)—are all linked to their intelligence.

Primate behavior and ecology are correlated with anatomy. For example, primate diet is linked to the shape of the teeth and skull which process food. Then body size and brain size are both linked together and both depend on the quality and amount of food. Adaptations for types of primate locomotion are evident in the chest, limbs, hands and feet, and tail. Primates manipulate and assess their environment with their hands, with stereoscopic and color vision, and with their intelligence.

Clues to primate mating systems come from sexually dimorphic traits like canine and body size. Amount of primate social complexity may be correlated to brain size as well since complex social networks require complex neural networks to keep track of them. Home-range size depends on group size, individual animal size, food distribution, size of the habitat, and competition with other groups (within the same species or not).

Types of social organization can cross superfamily boundaries within the primates. For instance, pottos (a type of slow lorise) and orangutans (a great ape) are solitary. All other types of social organization are gregarious, where adults aggregate in groups, and these include one male–multi-female groups (gorillas) and multi-male–multi-female (baboons and chimpanzees).

MONKEYS AND APES

Monkeys are divided into geographic groups, New World (The Americas) and Old World (Europe, Asia, Africa). Humans belong to the infraorder Catarrhini, which includes Old World monkeys (Superfamily Cercopithecoidea) and apes (Superfamily Hominoidea).

Catarrhines have a 2:1:2:3 dental formula which means that per quadrant of the mouth (there are four: upper right jaw, upper left jaw, lower right jaw, lower left jaw) adults have two incisors, one canine, two premolars, and three molars. Baboons (*Papio*) and macaques (*Macaca*) dominate the Old World monkey group with their diversity and their impressive geographic coverage (baboons extend from north to South Africa and macaques cover much of south, east, and Southeast Asia and even reach to Western Europe).

New World monkeys (infraorder Platyrrhini) have one more premolar per quadrant than the catarrhines (2:1:3:3), which is the primitive condition. They also have grasping prehensile tails, some of which are furless on the underside with the equivalent of fingerprints for better grip. The New World monkeys also tend to be more acrobatic in the trees than Old World monkeys. Types of New World monkeys include capuchins, howlers, spider monkeys, squirrel monkeys, marmosets, and tamarins.

No living primate eats strictly leaves, fruits, or insects, but that does not prohibit them from evolving special dietary adaptations. Leaf-eating monkeys (colobines) have a long gut and a sacculated stomach for digesting cellulose, similar to the way cows digest grass. Their teeth are also adapted to shear leaves with sharp crests on the molars in a condition called bilophodonty. Colobines, are more sedentary than other monkeys and have smaller brain size relative to body size, which is expected given their low-quality diet.

Gibbons and siamangs (family Hylobatidae) from Indonesia and South Asia are the most suspensory of all primates and are known as the "lesser apes." They are also the least sexually dimorphic of the hominoids. They move through the trees almost exclusively by two-armed brachiation and by four-legged climbing in between brachiating bouts. When on the ground, they walk bipedally, but it is rare. They have very long arms and hands and very short legs and feet (less weight on the bottom for better swinging through the trees). Gibbons and siamangs form monogamous mated pairs, so they have little sexual dimorphism as a result. Like Old World monkeys, they also have ischial callosities, which is an adaptation for sitting that includes roughened pads on the skin near the buttocks, and thickened bone underneath, but the other apes do not. The majority of the gibbon diet is fruit (figs) and they can weigh between 4 and 13 kg (between 9 and 29 lbs) depending on the species, and males and females average the same size.

Thanks to pioneering fieldwork by primatologists Jane Goodall (chimpanzees), Dian Fossey (gorillas), and Biruté Galdikas (orangutans), incredible humanlike details about great ape behavior and biology are constantly forcing us to reevaluate what it means to be human. These women immersed themselves into ape life, getting up close and personal with the creatures and set the precedent for researchers and conservation efforts to follow.

The "great apes" are much larger than gibbons and siamangs and include the orangutans (the only Asian great ape), gorillas, chimpanzees, and bonobos. Unlike the lesser apes, they are not monogamous so their anatomy differs accordingly.

Orangutans (*Pongo pygmaeus*) are also known as the red ape and live quiet and mostly solitary lives in the diminishing forests on the islands of Borneo and Sumatra. Their reproductive strategy is known as a dispersed polygyny and they have the sexually dimorphic bodies to match. Males can weigh up to 200 lbs but females weigh less than half of that. Females and their offspring defend territories from other females, while males attempt to secure mating privileges by controlling several different female territories. Large dominant males have wide cheeks and also use their throat sacs to make loud roaring songs. Orangutans are mostly arboreal and eat fruits from the trees but sometimes males get too large to be completely arboreal and must spend more time on the ground.

The African great apes, existing mostly in regions of west and central Africa, are the most terrestrial of the hominoids (excluding humans which are the most terrestrial hominoid of all). African great apes include *Gorilla gorilla* (subspecies include the "western lowland," the "eastern lowland," and the highly endangered "mountain" gorilla) from Cameroon to the Virunga Mountains along the border of the Congo, Rwanda, and Uganda; *Pan paniscus* (the bonobo, formerly known as the "pygmy" chimpanzee) from the central Congo; and *Pan troglodytes* (the "common" chimpanzee) from west, central, and east Africa, mainly Tanzania, Ivory Coast, and Uganda. When walking quadrupedally, they all walk on the knuckles of the phalanges of their hands both on the ground and on horizontal tree branches.

With the largest body size, gorillas are the most terrestrial. Gorillas often feed on vegetation on the ground, as opposed to exploiting food sources like fruits and nuts in the trees like the smaller bodied chimpanzees and bonobos. As the largest living primates, gorilla males can weigh up to 400 pounds (200 kg) and females average half the size of males. They are gentle, social animals.

Chimpanzees and bonobos, however, are much livelier. Chimpanzees are much less sexually dimorphic with males weighing about 150 pounds (68 kg) and females about 120 pounds (56 kg). Bonobos are called "pygmy" chimpanzees because of their slightly smaller and slender bodies compared to common chimpanzees.

In large part due to body size, gorillas nest on the ground, as opposed to chimpanzees and bonobos who build sleeping nests in trees. When moving about arboreally, gorillas, bonobos, and chimpanzees use both quadrupedal and suspensory (hanging) locomotor behaviors. Although slow deliberation characterizes the majority of gorilla climbing bouts, even the largest ones are capable of very fast high-energy arm-swinging, like the displays or fleeing behaviors of chimpanzees and bonobos. The frequency of arboreal behaviors, especially for gorillas, depends upon the size of the trees that accommodate them. Habitats for mountain and lowland gorillas are distinct enough to warrant anatomical differences between the subspecies, because the montane habitats of mountain gorillas offer fewer climbable trees compared to the rainforests that lowland gorillas inhabit.

The reproductive strategies of gorillas and chimpanzees differ. Gorillas mate within a one or two male polygyny where one or two males have sole access to a group of females. Such a system is beneficial to both the males and the females so the term "harem" which comes with a stigma associated with female oppression is no longer applied to gorilla reproductive behaviors. Males benefit from knowing paternity and females benefit from the protection of the male from other males that may harm their offspring. Chimpanzees on the other hand have a much more fluid mating strategy. Their social units are called fission-fusion because groups join and then disperse from one another and then rejoin regularly. Under such conditions matings can take place between multiple different male–female pairings and social hierarchies within groups can determine who mates with whom. Like most primates and all the great apes, chimpanzee offspring remain with their mothers, not their fathers, throughout development.

All primates evolved from a common ancestor with fish and although it is tempting to assume humans were the only to become fishermen, chimpanzees are known to fish for termites (see Chapter 5). Still, humans have a fair share of differences from their closest chimpanzee relatives, apart from what prey items they choose to fish for, and these traits will be discussed further in the later chapters.

3

Prehistoric Evidence

WHAT IS A FOSSIL?

Although they are usually hard to find, fossils are not hard to see. What is surprising to many first-time fossil hunters is how life-like a fossilized animal or plant appears. There is no need to use your imagination to spot fossils in the ground, as if conjuring animal-shaped clouds in the sky. Biological organisms are symmetrical, mathematical, patterned, and in most cases they fossilize having retained much of the appearance they had in life, even if they become flat and resemble prehistoric roadkill after millions of years of geologic pressure.

Depending on where one looks, fossils are sometimes harder to track than trophy game, yet in a few exceptional places, like the badlands around Lake Turkana, Kenya, it is difficult to step without crushing fossil fragments beneath your boots (Figure 3.1).

Fossils are the mineralized remains of once living organisms. Mineralization of organic material occurs through ground water replacement or through the leeching of minerals between the fossils and the surrounding earth. Chemicals will often replace the inorganic elements in bones and teeth as well. Fossils can be formed from things like bones, teeth, plants, wood, hair, feathers, footprints (like the bipedal walkway at Laetoli, Tanzania, about 3.6 Mya), tracks, trails, body impressions, feces, and vomit. Soft tissue, like skin, rarely fossilizes, but it is possible and it usually takes the form of an impression or "trace fossil."

When very young fossils do not turn entirely to rock they are called "subfossils." These include the skeletal remains of animals that have recently gone extinct like many species of lemurs on Madagascar, some of which were lost only within the last 200 years. Subfossils should not be

Figure 3.1 Paleoanthropologists survey the badlands of Koobi Fora, Kenya to search for fossils and sites. *Photograph by Holly Dunsworth.*

confused with "living fossils," which are organisms that have not changed much through time, like crocodiles that look very much like fossils of their 80-million-year-old ancestors.

Fossils are often hard to find because of the slim chance that a dead organism actually gets preserved and then the even slimmer chance that anyone will actually find it millions of years later (Figure 3.2). If a person wants to become a fossil, they need to make sure they are buried quickly before scavengers chew up and scatter their bones. They should plan to die near a river, pond, or lake or amongst sand dunes where water and air are constantly building up sediment. The sediment cannot be too acidic or contain too many microorganisms since these promote skeletal degradation. Very dry, arid places are best, but volcanic ash burial is the best scenario since, depending on the geochemistry of the ejected sediment, a person could meet the requirements of becoming a fossil, plus have the potential to be dated well by geochronology. Of course, the final requirement of becoming a fossil is to actually be discovered by someone, so someone would need to draw a map of their whereabouts using symbols that will be understood in a million years (when contemporary languages will have evolved beyond recognition), on a material that will also be preserved for eons.

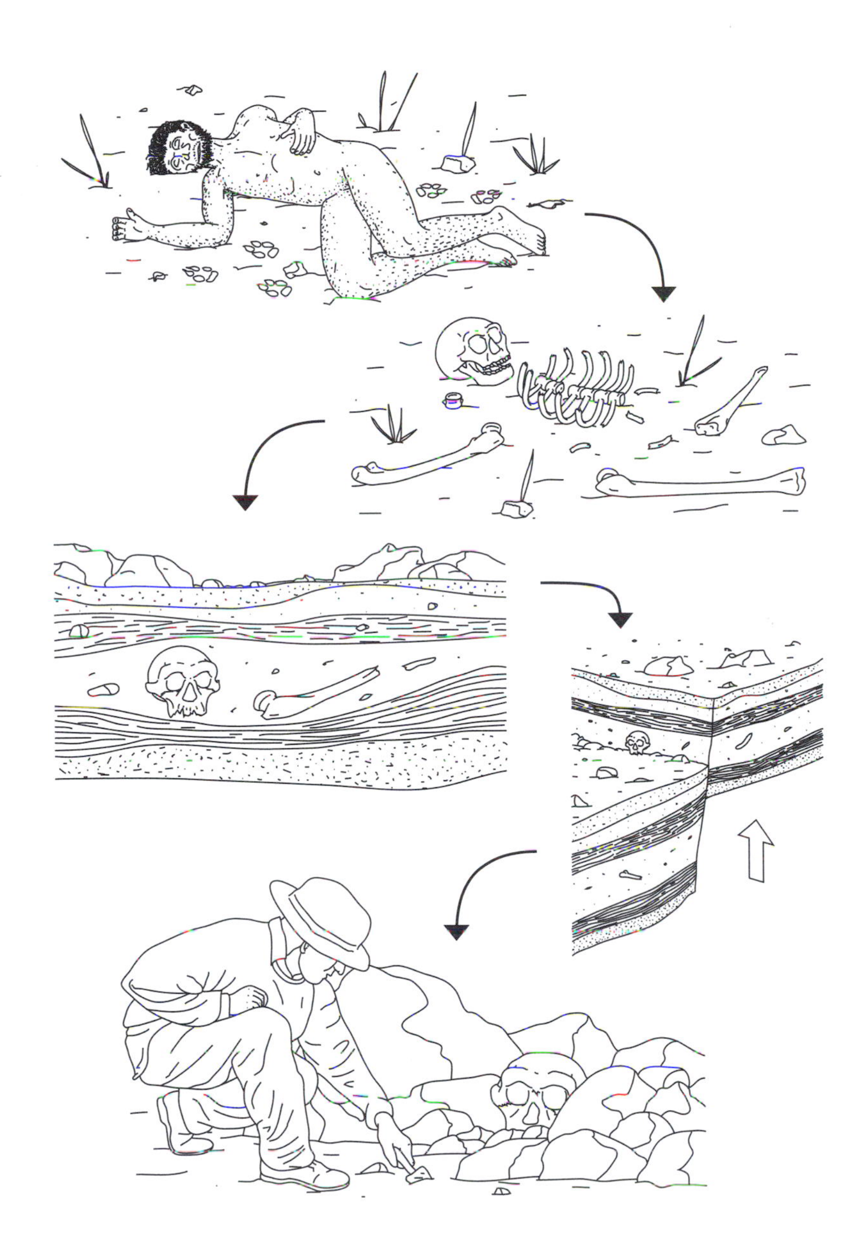

Figure 3.2 The Life of a Fossil. A *H. erectus* dies on the savannah and may or may not be devoured by scavengers like carnivores and birds. Next, particularly if the environment is an arid one, the body will decompose to leave just the skeleton behind and the bones may get scattered by animals and the elements. Through time, the bones get buried naturally by sediment carried by water or air. Geochemical processes and pressure under the ground turn the bone into rock. Then geologic processes like faulting (shown here) or simple erosion can expose the fossil at the surface for a paleontologist to discover. The paleontologist would excavate around the spot in order to find any other remains of the skeleton that are, hopefully, preserved there. *Illustration by Jeff Dixon.*

GEOLOGY AND DATING METHODS

Before mounting an expedition, paleontologists first look to geologic maps of a region to determine if the right time period is actually present. Then, ideally, they look at aerial photographs or they make a visit by airplane, automobile, or foot to determine if those rocks are exposed and accessible and free of thick vegetation or water cover. Often paleontologists and geologists will explore regions of interest together. If they are interested in fossil hominins they look at rocks dating to the Miocene, Pliocene, and Pleistocene. But if they spot enormous bones spilling out of Cretaceous rock layers, they will alert their colleagues who study dinosaurs.

Once fossils and artifacts are discovered, more precise dates need to be estimated for them in order to put them into the context of human evolution. Either the fossils and artifacts themselves are dated (direct dating) or the rocks and sediment that contain the fossils are dated (indirect dating). It is often best to date the rocks rather than the fossils or artifacts to avoid damaging the specimens.

Because of error ranges associated with dating methods, a combination of methods is used. By combining relative methods (approximate ages) with absolute methods (exact ages with error ranges) it is possible to put a specimen into a more precise context.

Relative methods can only indicate whether or not a fossil is older or younger than something else. They do not result in exact dates, but the principles are based on the three laws of stratigraphy:

1. *The Law of Superposition*—Deposits deeper in the ground are older than those that are closer to the surface.
2. *The Law of Original Horizontality*—Strata are deposited (e.g., by air, water, volcanic eruption) horizontally and if they do not appear horizontal, then geologic processes like faulting or doming have altered them.
3. *The Law of Original Continuity*—Equivalent deposits that are separated by a canyon, for example, were the result of the same depositional event and once belonged to the same continuous layer before erosion separated them.

During long stretches of the Earth's history, the magnetic field was opposite to what it is now. The present state is called "normal" and the opposite is known as "reversed." Iron crystals in heated rocks will act like magnets and orient themselves in the direction of the poles when the rock cools. So the nature of the Earth's magnetic field at the time the rock formed is recorded in its polarity. Rocks that are dated absolutely and that also reveal their polarity allow geologists to construct a sequence of paleomagnetic changes through time and these are verified by the record known from deep sea cores. Then if a fossil is found in a rock that does not allow absolute dating, the geologist can apply relative-dating logic to the paleomagnetic readout (based on whether the rock is "normal" or "reversed") to place the fossil in a particular time frame in history.

Paleoanthropologists find many more fossils of nonhominins than hominins and these allow them to reconstruct the environment that hominins were living in. Nonhominin fossils are also helpful for relative dating with a technique called biostratigraphy. The age of a hominin fossil can be determined by the fossils of other well-known (and well-dated) animals that are associated with it. For example, if a certain well-known species of fossil pig is found in the same layer as a hominin fossil at site A, and this pig species is also found at site B that is dated to 6 Mya, then it is probable that that site A is also 6 Mya and therefore the hominin fossils are that age too.

How was the pig at site B dated in the first place? Absolute dating is a way to estimate dates for fossils and artifacts, usually within a range of years. Methods of radiometric dating are a large component of the absolute dating approach. Radiometric dating methods are based on the known constant rate of decay of isotopes (radioactive forms of elements) through the loss of particles in the nucleus of atoms. Radiocarbon dating, or carbon dating, is the best known of these methods. Since all living organisms contain carbon, artifacts like wood and bones can be directly dated with this method. Plants and animals take in C^{14}, an isotope of carbon throughout their lives. When they die, the C^{14} begins to decay into nitrogen (N^{14}). Death starts the radiometric clock ticking. The decay of C^{14} happens at a constant known rate and when the C^{14} present in an old bone is measured, an age can be given to it based on how much has decayed. The time it takes for half of the radioactive isotope in a sample to break down is the half-life. Half-lives are different for every radioactive isotope. The half-life for C^{14} is 5,730 years, so that in that time, half of the sample of C^{14} has decayed. With such a short half-life, specimens over 50,000 years cannot be dated according to C^{14} decay because in that time all the C^{14} will be decayed and gone from a specimen, making it impossible to measure and date. Hominin fossils from Europe and Australia have been dated with this method, but the applicable scope of carbon dating excludes most of the fossil record for human evolution.

The half-life of potassium (K^{40}) is 1.3 billion years. It decays much slower than C^{14} so dating methods using potassium are applicable to the rest of the hominin fossil record over 100 Kya. Garniss Curtis and Jack Evernden developed a method which tracks the decay of K^{40} into argon (Ar^{40}) in rocks, a method that is similar to carbon-dating except instead of dating the object of interest it dates the rocks around the object.

During the extreme heating that occurs during volcanic eruptions, argon is eliminated from the rocks in the volcano. When the lava cools, it contains the radioactive isotope K^{40}, but no Ar^{40}. The cooling and solidification of the volcanic rock sets the clock at zero and starts it (just like when the organism dies for carbon dating). K^{40} decays into stable Ar^{40} within the solidified lava rock over time. If that rock happens to contain a hominin fossil or if it is located just above or just below a hominin fossil or stone tool, a sample of it will be collected and then taken to the laboratory to be heated. Scientists will measure the relative amounts of K^{40} and Ar^{40} to determine how much decay has occurred and based on that, they can estimate how old the rock is and therefore how old the fossil associated with it is. A somewhat more precise variation of the K/Ar method is based on the same idea and measures the amount of decay of Ar^{40} into Ar^{39}, in a method called 40Ar/39Ar.

Obviously K/Ar dating does not work for hominin sites without volcanoes nearby to spew ash and lava. Luckily many of the hominin fossil sites in East Africa are located in proximity to the Great Rift Valley, which is a volcanically and geologically active fault zone where Africa is literally splitting apart. The Rift's many volcanoes were highly active during the course of hominin evolution, laying down layers of ash called "tuffs" that enable geologists to build a dating sequence (Figure 3.3). For example, fossils preserved between ash A dated to 1.6 Mya and ash B dated to 1.5 Mya are estimated to be between 1.6 and 1.5 Mya, or averaged to 1.55 Mya. Not only are tuffs convenient for absolute dating, but because each tuff from a volcanic eruption has its own unique chemical fingerprint, tuffs can be traced over great distances to allow sites to be correlated and compared.

Hominin sites between the ranges of radiocarbon and K/Ar dating are dated with techniques like electron spin resonance (ESR) and uranium-series dating (based on uranium isotope decay) and these work best between 500 and 50 Kya. ESR depends on the measurability of radiation that is given off when isotopes decay. Thermoluminescence (TL) is also based on radiation that is emitted, but estimates the amount of radiation by heating the rock ("thermo") and measuring the amount of light given

Figure 3.3 A geologist searches a site in the Middle Ledi region of Ethiopia (near the Hadar region) for the right type of volcanic ash to use for 40Ar/39Ar dating. The dark bands of volcanic ash are approximately 3 million years old and are preserved within layers of white lake deposits. *Photograph courtesy of Guillaume Dupont-Nivet and Chris Campisano.*

off ("luminescence"). TL is the preferred method for dating hearths at home bases and camp sites since it is possible to decipher if and when a rock was heated (that is, "reheated" since it was first heated during its formation).

Different dating methods are appropriate for different materials of differing ages. Whenever possible, scientists use multiple methods. Unlike radiocarbon dating, however, which can sometimes date an artifact to within 100 years, methods for hominin fossils rarely get any closer than estimated ranges within a few thousand years. With increasing technology, the resolution of dating techniques is getting much more precise, with error ranges becoming smaller and smaller. The ability to precisely pinpoint a fossil's place in time in the story of hominin evolution (within a few thousand years as opposed to a few tens of thousands of years) is becoming the expectation.

CLIMATE CHANGE AND PALEOENVIRONMENT

Hominin evolution took place during a time of great global climate change. Drilled-out deep sea and glacier cores hold historical records of the Earth's climate. Each layer of ice that formed, and each layer of sea floor that was deposited, contains a snapshot of the climatic conditions during its creation. Those conditions are told by the amount of certain chemical compounds in the layers. Times of cooling (glacials) or warming (interglacials) are indicated by ratios of oxygen isotopes differentially taken up by microscopic, one-celled organisms (foraminifera or "forams") that are fossilized in the layers.

Some patterns suggest that extinction rather than speciation events are more likely to be correlated with climate change. However, extinction resulting from such changes can open up niches for new organisms to fill. The extinction of the dinosaurs at the Cretaceous-Tertiary boundary was gradually followed by the enlargement of mammals that eventually filled the empty small- and large-bodied dinosaur niches.

The evolution of the earliest primates immediately predated the rapid global cooling that began in the Eocene. Another long-term cooling and drying trend coincided with the origin of hominins between 8 and 5 Mya. Oscillations in global climate occurred during hominin evolution and these oscillations probably affected the outcome of human evolution. Starting in the late Pliocene around 2.8 Mya Africa became cooler and dryer, which likely converted forests to grasslands. East African vegetation transitioned from closed forested canopies to open, arid savannahs with reduced and seasonal precipitation.

Then continuing through the lower Pleistocene (1.8 Mya–900 Kya) the world climate began to cool more rapidly (Figure 3.4). The middle Pleistocene (900–125 Kya) was characterized by cold glacial periods, for 100,000-year cycles, interspersed with warmer interglacial periods. And this was followed by the "Ice Ages" of the Upper Pleistocene (125–12 Kya), which were also very cool glacial periods that were broken up by warm periods. During the colder periods, water would have been locked up and frozen in glaciers, causing sea levels to lower. Lower sea levels probably had a great impact on hominin mobility since isolated islands or distant continents could become attached to the mainland by dry land. Then when the ice caps melted in a warm interglacial phase, those regions could become islands again and isolate the species living on them.

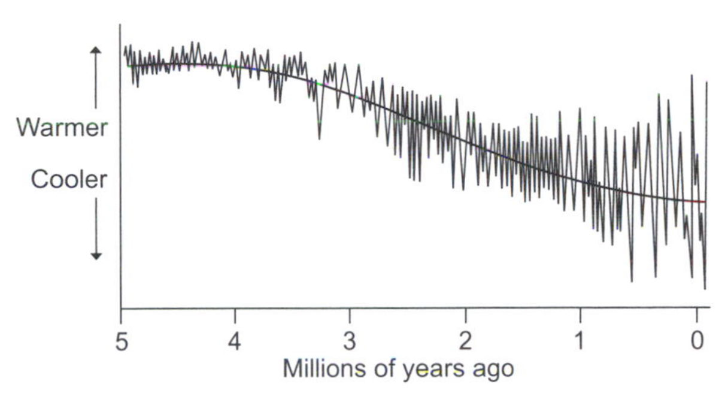

Figure 3.4 Global temperature, as measured by ratios of oxygen isotopes in deep sea and ice cores, has undergone a steady average decrease in the last 5 million years marked by huge fluctuations. *Illustration by Jeff Dixon.*

Fossils of animals and plants at a hominin site are useful for reconstructing the paleoenvironment in ways that differ from the tiny forams in sea cores. Based on what we know about the diverse ecology of living bovid species (antelopes and the like) the types of fossil bovids at a site can indicate whether it was open grassland or more closed woodland. Fossil monkeys are also reliable indicators of environment: if they have arboreal adaptations in the skeleton or teeth that are associated with leaf-eating (like modern colobines), then the environment was probably forested. If they have terrestrial adaptations like baboons then the hominin site may be much more open, arid, and savannah-like. Often crocodile teeth are scattered everywhere at sites near ancient watering holes or ancient riverbeds. The more animals a paleontologist can use to reconstruct the paleoenvironment, the stronger the science.

THE EARLIEST PRIMATES AND FOSSIL MONKEYS

The mammalian fossil record is full of fascinating extinct animals, a subset of which is comprised of primate fossils. The earliest primates are found in the Paleocene and resemble squirrels. Primates split from the rest of the mammals very early along with rodents, so fossil primates are some of the oldest and most primitive fossil mammals on record.

The earliest animals to resemble primates are called plesiadapiforms (family Plesiadapiformidae) and are found in rocks in Europe and western North America (particularly Wyoming). The rock composition and the related fossil fauna indicate plesiadapiforms preferred warm wet forests that were dominant during this period in this region.

Like most early mammals, plesiadapiforms were nocturnal quadrupeds with good senses of smell—a trait inferred by large snouts and comparatively small eye sockets, which indicate relatively less reliance on vision. Like modern primates, plesiadapiforms had characteristic teeth and grasping hands and feet. But unlike modern primates they had claws, they lacked leaping traits of the skeleton (primitive for primates), and they did not have large, enclosed bony eye sockets.

Angiosperms, or plants that disperse seeds with flowers and fruits, evolved in the Cretaceous period creating a new food source for animals and a niche that the earliest primates capitalized on. They were arboreal, fruit, seed, and insect-eaters. There is a debate over which food item—animated insects (many of which were pollinating the angiosperms) or colorful fruit—was the driving force in evolution of primate features like acute color and stereoscopic vision and dexterous grasping hands. Although it could have been both insects and fruit since all modern primates eat a variety of foods, and are not limited to one specialization.

By the Eocene, the fossils of the first fully recognized primates that display the primate features plesiadapiforms lacked, lived in North America and Europe which were connected by land at the time. The omomyids (from the family Omomyidae) were small insectivores resembling modern tarsiers. The adapids (from the family Adapidae) were larger fruit- and leaf-eaters resembling modern lemurs.

If we were lemurs, we would omit the following sections on fossil monkeys and apes and humans and we would follow the fossil trail of our ancestors from the adapids through to our present-day life on Madagascar. But we are humans, so we will follow the fossil trail of our ancestors from the earliest primates to monkeys, then to apes, and finally to humans (which includes those much larger, bipedal primates that also live on Madagascar).

All monkeys, apes, and humans (haplorhines) share a common ancestor from Asia in the early Eocene known as *Eosimius* or "dawn monkey."

Eosimius is remarkably tiny, as small as a person's thumb, but has features that distinguish it from lorises and lemurs and link it to monkeys.

Much of what we know about early haplorhine evolution comes from fossils preserved in the Fayum desert of Egypt—some 60 miles southwest of Cairo. Sites there were first excavated in the early part of the 20th century and continue to be excavated by teams led by Elwyn Simons. These sites on the edge of the Sahara desert are very dry today, but they are rich in fossil mammals from a time when the region was a forested swampland, 34 Mya. The primitive monkeys *Apidium* and *Aegyptopithecus* hail from the Fayum. *Apidium* has the dental formula of platyrrhines (New World monkeys), but *Aegyptopithecus* has that of catarrhines (Old World monkeys, apes, and humans). Both species have fused mandibles and totally enclosed bony eye sockets. The fossil record for Old World monkeys, once they split from the ape and human lineage, is very rich by the Pliocene and Pleistocene but earlier on in the lineage's evolution, fossils are rare.

Explaining how monkeys made it to South America is a big challenge. New World monkeys evolved in isolation from a common ancestor with Old World monkeys and since then they have become very diverse. Fossil New World monkeys first appear in South America at about 28 Mya. Some of the earliest fossils, found in Bolivia, are very similar to those discovered in the Fayum from the Oligocene. There are two possible routes to South America but both seem difficult. One path is from North America, but the fossil record tracking any sort of monkey evolution throughout North America at this time is very sparse and furthermore, the isthmus at Panama was not connected to South America yet, so they would have had to island-hop through the Caribbean. The second possible route would have been across the south Atlantic ocean from Africa where the early fossil precursors are found. New World monkeys may have floated over on rafts of clumped vegetation or "floating islands." The two continents were closer to one another because sea floor spreading at the mid-Atlantic ridge was in an earlier phase of pushing the Americas away from Europe and Africa (which is still happening at the same speed your fingernails grow).

FOSSIL APES

At present monkeys far outnumber the apes, but back in the Miocene, the situation was reversed. There are comparatively few monkey fossils in the Miocene. At that time, apes had radiated to fill most niches that monkeys later took over and fill today. The ape diversity decreased through time so that by the Pliocene, monkeys were dominant and there is a rich monkey fossil record from around the Old World to prove it. Possibly, monkeys simply outbred apes. Great ape mothers wait over four years between births because their offspring develop much slower than monkeys.

Most of the fossil record for apes hails from East Africa and Europe with a few species found in Asia and Indonesia. Early apes show an increase in body size and a diversity of locomotor modes from those of their monkey-like ancestors. Although there are numerous Miocene apes on record, not one is confidently placed on the direct lineage leading to humans. The mosaic nature of the apes, that is their mixture of traits that makes them unlike any living apes today, creates a challenge for paleontologists to interpret them.

During the Miocene, East Africa was covered in rain forests and filled with apes of a broad variety of sizes and diets. These fossil species shared trends with living apes: large brains, fewer vertebrae, long upper limbs, and no tails. Their skulls and teeth show they ate fibrous foods like fruits, nuts, and tough vegetation. They had thick tooth enamel, rounded cusps, flared cheekbones for jaw muscles (*temporalis*) to fit underneath. Postcranial adaptations of these apes included anatomy built for arboreality, climbing, and brachiation.

Only one Miocene ape has an agreed-upon ancestor–descendent relationship with a living ape and that is *Sivapithecus* from 14 Mya in the Siwalik Hills of Pakistan. *Sivapithecus* is an ancestor of modern orangutans. There are many candidates, but as of now here is no consensus as to which one(s) is the ancestor of the African great apes: gorillas, chimpanzees, and humans.

The best-known genus of Miocene ape is *Proconsul* known mostly from sites on Rusinga Island in the Kenyan waters of Lake Victoria, dating from 20 to 18 Mya (Figure 3.5). There are over twelve partial skeletons of *Proconsul* and several jaws, isolated bones and teeth, and one well-preserved skull. *Proconsul* is best imagined as having a monkey-like body with ape-like traits in the teeth (fruit-eating adaptations), skull, some aspects of the limbs, and lacking a tail. Certainly the lack of evidence for something, especially in the fickle fossil record, does not prove it does not exist. However, none of the partial skeletons have preserved caudal (tail) vertebrae and the sacrum does not appear to have any anatomical ties to a tail. Since it is assumed that tails were lost only once in hominoid evolution, the lack of tail in *Proconsul* lends support to its ape status.

Numerous Miocene apes with ape-like features could have been direct ancestors to living apes. *Morotopithecus* from 21 Mya in Uganda is

Figure 3.5 Two 18 Mya feet of the stem ape *Proconsul* were discovered still in anatomical articulation at the Kaswanga Primate Site on Rusinga Island, Kenya. *Photograph by Mark Teaford.*

interpreted to have brachiating and suspensory characteristics like the great apes. As fossil apes approach the middle Miocene in age, they take on many more modern characteristics. The African fossil apes *Nacholapithecus, Kenyapithecus, Kamoyapithecus, Samburupithecus, Otavipithecus,* and *Afropithecus* are all contenders for African ape ancestors. But there are even more ape-like species in Europe. Around 12 Mya *Dryopithecus* emerges in Spain, Romania, France, and Hungary. *Oreopithecus* is so ape-like, it has been interpreted to even have adaptations for incipient bipedalism, but those notions are not strongly supported. Other European fossil apes include *Ouranopithecus* (Greece), *Ankarapithecus* (Turkey), *Griphopithecus,* and *Grecopithecus.* The most recent find, a well-preserved partial skeleton of *Pierolapithecus* shares many traits with living great apes. The Miocene ape record is puzzling. It is assumed that since all but one of the great apes live in Africa that they evolved there, but the fossil apes from Africa are more of a mosaic of monkey-like features than their less monkey-like, more ape-like relatives in Europe. After more fossils are found in the late Miocene of both Africa and Europe, the story of ape evolution and the emergence of the hominin lineage will be much clearer.

Gigantopithecus: A Real Bigfoot

Perhaps the legend of Bigfoot has such staying power because of its ancient roots. The Pleistocene fossil record of China contains teeth and jaws of a prehistoric Bigfoot called *Gigantopithecus blacki.* Although no skeletons (or foot bones) have been discovered, the estimated body size based on the size of the teeth is enormous, at over 400 kg (880 lbs), and that is a conservative estimate. Even if the teeth were larger than expected for body size, *Gigantopithecus* was still the largest known primate to ever exist. Unlike the legendary Bigfoot, *Gigantopithecus* would have probably walked quadrupedally like another comparable legendary character, King Kong. The giant ape only went extinct about 100 Kya, so it had plenty of temporal and geographic overlap with *Homo erectus.*

Gorillas have humanlike feet because adaptations for carrying their large heavy bodies and for spending time on the ground actually mimic adaptations in human feet that accommodate for bipedalism and terrestrialism. As a consequence, gorilla footprints also look similar to those of humans. If *Gigantopithecus* was anything like modern gorillas, which it probably was, its feet could have been gorilla-like and maybe even humanlike. Perhaps the Bigfoot trackers of the world should trek to China to find fossilized footprints of a genuinely real, albeit ancient Bigfoot.

BUSHES AND TREES

Until fossils come with tags that describe exactly what species they belonged to, what they ate, who they mated with, and whether or not they could speak, there will always be arguments over their interpretation. The issue is not whether fossil hominins hold clues to our evolution, the issue is how to interpret those clues and come to a logical consensus on how *Homo sapiens* came to be.

The first obstacle in coming to such a consensus is that hominin fossils are rare. Sure, it is hard to visit any part of the world without running into humans today, but our insect-like colonization of Earth is a very recent phenomenon. Prior to 10 Kya all humans lived in small groups rarely exceeding 150 individuals. So the chances of finding a hominin fossil are diminished purely because of past population sizes. Paleoanthropologists find many more bones and teeth of rodents and bovids (antelopes and the like) than they find hominins, but there are still thousands of hominin fossils on record.

Because of two fundamentally different views of interpreting evolution, there are arguments over which fossils should be included in the lineage leading to humans and which should be placed on side branches. Basically, the "splitters" see the hominin phylogeny as a bush, but the "lumpers" see it as a tree. The most extreme lumpers see it as a saguaro cactus, with maybe two or three lineages/branches at most.

The bushy view provides a very complicated phylogenetic history of hominins with multiple species existing at any one time. The tree or cactus symbolizes the view that hominins have evolved with very little diversity and with very little overlap in species, in long evolutionary lineages.

The splitters argue that the fossil record cannot possibly contain representatives from all of the hominin species that ever existed; that there must be so much diversity that is not being sampled because of preservation issues. The lumpers' view is a conservative one that points to evidence from the fossil record as we know it.

The bush-versus-tree argument is important if we are to determine the nature of human evolution. It is also important for how we reconstruct the daily lives of hominins 1.5 million years ago. Life may be a whole lot different if there was a separate species of 3-foot-tall bipeds sharing the planet with us, and that was what life was like for humanlike *H. erectus* which coexisted with ape-like robust australopiths and for Indonesian humans about 13,000 years ago when the so-called "hobbits" lived on

Flores. What if being the only bipedal tool-using ape, like we are now, is an exception rather than the rule?

Why "Hominin"?

"Hominin" is replacing the once ubiquitous term "hominid" in the literature. Both terms are used for humans and their extinct ancestors since the last common ancestor with chimpanzees (LCA) and both continue to be accepted. The difference in the terms has to do with how scientists view the relationship between humans and the rest of the great apes. Traditionally humans were the only species to belong in the family Hominidae ("hominids"), with the great apes placed in a separate family, Pongidae, because anatomical similarities group the great apes to the exclusion of humans. However, due to the genetic similarity between chimpanzees, humans, and gorillas (to the exclusion of orangutans), chimpanzees, gorillas, and all their fossil ancestors are increasingly lumped in with humans in Hominidae. Therefore subcategories were created to further differentiate the groups, where humans and their fossil ancestors are called hominins.

THE LAST COMMON ANCESTOR

What would the last common ancestor look like? What would the earliest hominin look like? How much would it resemble a modern chimpanzee? Paleoanthropologists must show that their fossils are not fossil chimpanzees or gorillas in order for them to be accepted as part of the hominin family tree. This is a tall order for the early part of the hominin fossil record (between 7 and 4 Mya) when hominins were still very ape-like. Plus there are no fossil chimpanzees or gorillas from the late Miocene to the Pliocene to hold up for comparison.

There is only one recognized fossil chimpanzee on record and it dates to the relatively recent mid-Pleistocene. A few teeth were collected from a site near Lake Baringo, Kenya. Argon-Argon dating put the teeth at about 525 Kya. They look exactly like modern chimpanzee teeth and, interestingly, there are fossil humans from the same regional localities, which presumably lived in the same place in prehistory as the chimpanzees.

The tropical habitats of chimpanzees and gorillas are to blame for their near absence from the fossil record. Modern African apes stick to warm wet tropical forests and are assumed to have enjoyed the same habitats in the past. Fossilization rarely occurs anywhere, let alone in these kinds of habitats where soil acidity is high and where organisms that devour carcasses are thorough and efficient.

It is difficult to recognize and properly identify primitive species near major evolutionary divergence events, like the LCA and its relatives. On top of that, how do we gauge within and between species variation for extinct creatures? Basing variation estimates on what we see in monkeys, apes, and humans today is the most logical and reasonable way to overcome this problem, but it also risks masking different patterns of variation that may have existed in the past.

There are other obstacles in paleontology. Often the fossils are broken or fragmented. Some of the most diagnostic anatomy in the skull is also the most fragile. Knowledge from years of anatomical study and experience is required to piece together fragmentary bones but with the help of modern imaging techniques, like CT scanning, the process is becoming much easier. With virtual fossil reconstructions, paleontologists are able to repair fossil breaks without having to clean, extract, and assemble them by hand which often damages fossils even further.

The number of juvenile hominin fossils that are discovered can also lead to issues in their identification. Individuals that are not fully adult or fully grown do not have all of the diagnostic features of their species. A good proportion of important hominin specimens are infants or children, for instance, the Taung Child (*Australopithecus africanus*), the Nariokotome *H. erectus* boy, the Dikika baby (*Australopithecus afarensis*), the Mojokerto Child (*Homo erectus*), and there are several juvenile Neanderthals. Their ages are determined by their stage of dental eruption. Both the deciduous or milk teeth and the permanent second set of teeth erupt in a particular pattern and follow a particular schedule. For instance, the human first molar erupts around six years of age, and so forth. There is a similar pattern and schedule to the fusion of growth plates at the ends of the long bones. Differences in body size and muscle attachment size on the bones indicate whether or not the individual was male or female, but such differences are much less pronounced before adulthood so determining the sex of juveniles is difficult the younger they are.

Since fossils are relics of a creature's structure, species are identified by anatomical traits—mainly highly diagnostic features of the skull and teeth. For many of the terms used in the following anatomical discussions, please refer to the figures and to Appendix A and for the context of the species within the hominin phylogeny please refer to Figure 1.5.

Table 3.1 Trends through Time in the Plio-Pleistocene Hominin Fossil Record

- Reduction of face size
- Reduction in prognathism, or the projection of the face
- Reduction in molar size, relative and absolute
- Reduction in canine size
- Loss of arboreal characteristics like long curved fingers and toes, short legs, and long forearms
- Acquisition of terrestrial and bipedal characteristics like shorter toes, longer legs (especially the femur), and shorter arms
- Increase in cranial capacity (which is used to infer brain size)
- Increase in skull roundness
- Increase then decrease in browridges
- Increase in stature and body size
- Decrease in the robusticity of bones
- Decrease in sexual dimorphism, or the differences between males and females in body size and tooth and skull characteristics

In general there are trends in the hominin fossil record we can follow to track our ancestors' journey from LCA to us (Table 3.1).

QUADRUPEDAL TO BIPEDAL

Many of the changes that occur along the hominin lineage involve adapting to bipedalism from a quadrupedal ape's body (Figure 3.6). Built for climbing, swinging, and brachiating through the trees, ape bodies feature short, flat trunks, broad hips, long collarbones (clavicles), shoulder blades (scapulas) on the back instead of on the side of the body (like a monkey or a dog), a round head of the humerus which makes for a flexible rotator cuff at the shoulder, semierect posture, long arms, short legs, elongated forelimbs, long curved fingers, and no tail.

Humans still possess many ancestral ape-like traits that were the springboard for not only the evolution of bipedalism but also the evolution of overarm throwing and tool making which depends, in part, on our retention of some of this brachiating anatomy. Major changes in the limb proportions occurred due to bipedalism, since it is better to have long legs for a striding gait, humans developed much longer legs for the body size than apes have. We also have much shorter arms than apes relative to our body size and a few of the reasons are probably to do with greater control during tool use and throwing and also the optimal arm length for swinging arms by the side during walking and running.

The earliest hominins had a similar body plan to living gorillas, chimpanzees, and bonobos and it is debated whether or not they walked on their knuckles like them too. The knuckle-walking wrist must be strong and rigid to support the weight of the animal, so the morphology of the radius shows a locking mechanism to prevent collapse at the wrist. There may be anatomical remnants of knuckle-walking adaptations in early australopiths but there are few arm and wrist bones and the traits appear to be intermediate. Evidence from earlier hominins, with less bipedal adaptations (and presumably a higher chance of retaining such adaptations if they were there in the first place) has not been discovered yet.

Parsimony supports the LCA as a knuckle-walker. If the common ancestor of gorillas and chimpanzees was a knuckle-walker, then it makes sense that the LCA of humans and chimpanzees was one too, and that humans lost knuckle-walking adaptations as they evolved bipedalism. Nature does not always obey our rule of parsimony, so it is also possible that knuckle-walking arose independently on both the gorilla and chimpanzee lineages, meaning that it is possible the LCA did not yet have knuckle-walking behavior. Genetic analyses have shown that chimpanzees and humans are more closely related than chimpanzees are to gorillas so independent acquisition of knuckle-walking in the great apes cannot be ruled out yet. More fossils of ape and human ancestors from the crucial period during the late Miocene and early Pliocene will eventually sort out the hominin knuckle-walking question.

THE EARLIEST HOMININS

Until very recently, little pertinent fossil evidence was known from the late Miocene epoch when chimpanzee and human ancestry diverged. Now there are a few contenders for the title of earliest hominin—*Sahelanthropus*, *Orrorin*, and *Ardipithecus*—and there will need to be more fossil discoveries for paleoanthropologists to reconstruct the root of the hominin tree. Some questions paleoanthropologists ask about the earliest hominins include: Were they woodland creatures? If so, did bipedalism evolve in the forest as opposed to the arid savannah where it is assumed to have evolved? How many species of early hominins were there? Did the first split from the LCA occur by sympatric or allopatric speciation?

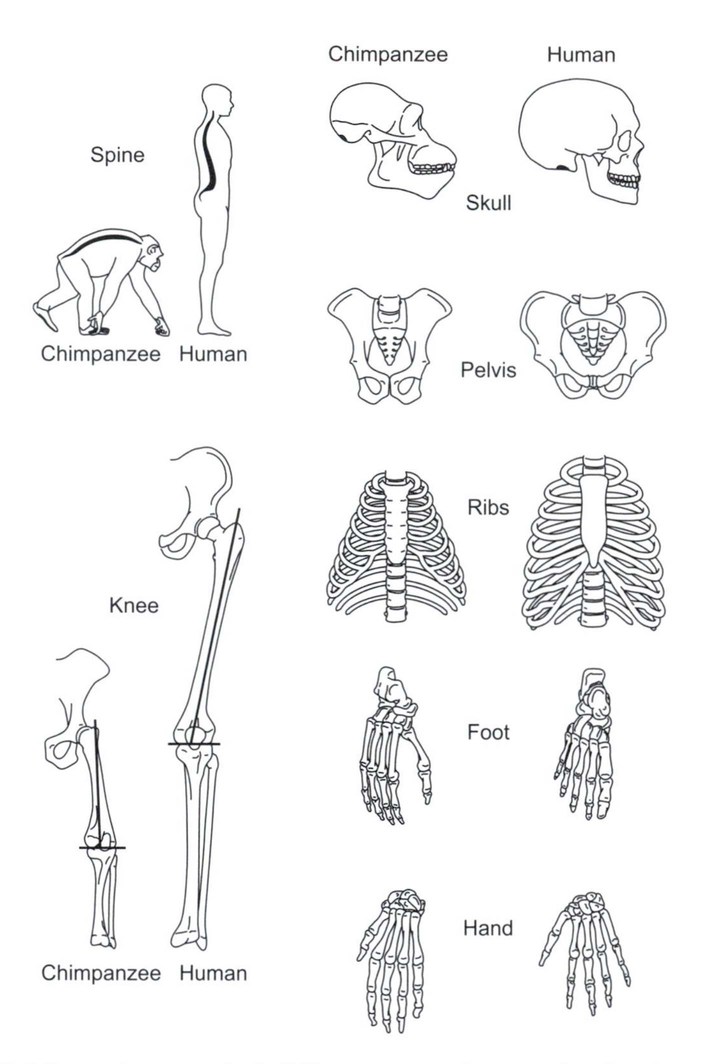

Figure 3.6 Several anatomical differences can be seen in the skeletons of chimpanzees and humans due to their completely different locomotor modes: quadrupedalism and bipedalism, respectively. These differences are used as clues for understanding the transition to bipedalism in the hominin fossil record. The human spine has curves at the upper back and neck and at the lower back to make it s-shaped as opposed to the chimpanzee back which is c-shaped. These curves in the human vertebral column help hold the head and torso above the pelvis and the center of gravity. The human knee is angled toward the middle to bring each leg directly under the center of gravity while walking and running. Because the chimpanzee knee is straight, the animal wobbles side to side when trained to walk bipedally. The skulls are not only different because of their brain sizes and face shapes, but because they are oriented differently on the body. There is a difference *(cont.)*

Sahelanthropus

French paleoanthropologist Michel Brunet is well known for exclaiming to the media, "I knew I would find it!" after he discovered the fossil he named *Sahelanthropus tchadensis.* It is a single cranium (skull without mandible) dating between 7 and 6 Mya. Found in the dry Lake Chad Basin near Toros-Menalla, *S. tchadensis* means "Sahel-ape from Chad." The fossil was nicknamed "Toumai" meaning "hope of life" in the local language. Thanks to molecular clocks that have estimated the LCA to have lived between 8 and 4 Mya, paleoanthropologists like Brunet who are searching for the earliest hominins are more confident about exploring sediments older than 5 Mya. The associated faunal remains indicate the region was once woodland, not at all like the sandy desert it is today.

Because there is no postcranial skeleton, the anatomy does not show direct evidence of bipedalism. However, it shares other characteristics with later hominins. The skull shows relatively small teeth, little prognathism, and has the foramen magnum under the skull (Figure 3.7). There is also no ape-like canine-premolar (CP3) honing-complex, which is a diastema (or gap) between the canine and premolar of the lower jaw to accommodate and sharpen the large canine of the upper jaw when the jaws are closed together. However, *Sahelanthropus* has ape-like teeth as well as many ape-like aspects of the skull. The cranial capacity

---

in the attachment of the vertebrae which encase the spinal cord. The spinal cord, thus, exits the skull toward the back of the chimpanzee head (see black spot) as opposed to exiting further underneath the human head. Since the torso is located atop the pelvis in humans instead of out in front like in chimpanzees, it serves as a basin for holding the organs. This basin-shaped pelvis anchors the muscles for walking and running (like the gluteus maximus muscles) in such a way as to allow them to provide much better balance. Bipedalism requires a great deal of balance since the body is only supported by one leg for much of the motion. Ribs are very different between quadrupeds, which have more of a funnel or conical shape, and bipeds, which are straight-sided and barrel-like. Because all the body weight of bipeds is shifted to the hindlimb, there are drastic changes in the feet. The human foot compromised its use as a grasping organ, like that of chimpanzees, in order for it to be a more sturdy and efficient locomotor platform. The most obvious difference is at the big toe which is thumb-like in chimpanzees. Freed from its locomotor role, the human hand has lost adaptations for quadrupedalism, tree-swinging, and climbing in favor of those for dexterity and manipulation of tools. *Illustration by Jeff Dixon.*

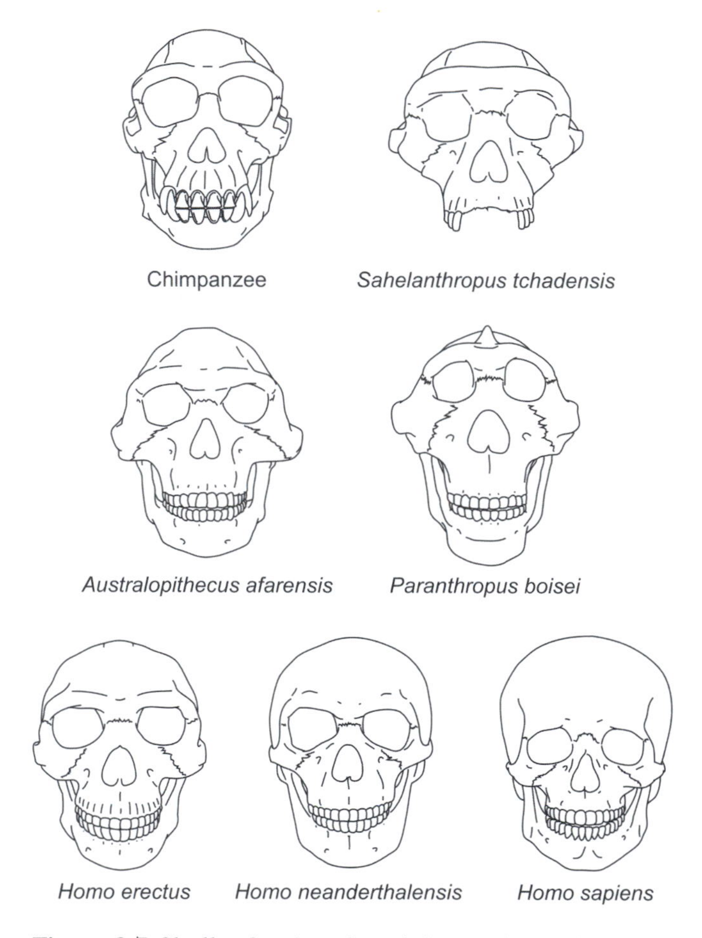

Figure 3.7 Skulls of various hominin species, with a modern chimpanzee and human for comparison, show their diverse features. There are no mandibles yet for *Sahelanthropus. Illustration by Jeff Dixon.*

at 350 cc is small, but not necessarily any different from what one would expect in the earliest hominin, nor from an early chimpanzee or gorilla ancestor.

The discoverers along with many other scientists consider *Sahelanthropus* to be an early hominin, which makes it the earliest specimen in the human lineage. But, there is criticism that because there is no skeleton to go with the skull, it is not yet known if this species was bipedal. Furthermore, the browridges are much bigger than expected. Browridges do not get large until after 2 Mya with *H. erectus* and earlier hominins do not have large browridges. Gorillas and chimpanzees, however, do have browridges and some critics argue that *Sahelanthropus* is more likely a fossil ape.

The dating of *Sahelanthropus* is also difficult to interpret since there are no volcanic sediments available in the sea of sand that is the Sahara Desert. There is only one species of pig that correlates to a dated site, Lothogam, in Kenya, and the biostratigraphic correlation points to an age closer to 6 Mya not 6.5 or 7 Mya. If *Sahelanthropus* is indeed 6 Mya, then it is very similar in age to the East African hominin *Orrorin tugenensis* (see next section). Unfortunately few of the preserved parts overlap for direct comparison, so more fossils are needed to understand the relationship better between these earliest hominins.

Orrorin

The earliest postcranial fossils of hominins are from 6 Mya. They include a couple of fragments of leg bones and a few teeth of *Orrorin tugenensis* from the Tugen Hills of the Lake Baringo region of central Kenya. Brigitte Senut and Martin Pickford made the discovery just around the turn of the last century, which inspired its nickname "Millennium Man."

Orrorin is humanlike in its thick tooth enamel, tooth size, and morphology. It has also been claimed to be bipedal because of morphology of the femur. The neck of the femur is long like in humans. Plus, x-ray images indicate the bony infrastructure is more like humans than apes. That is, the bone was built up to withstand the types of forces generated by walking upright on two legs. The bipedal status of *Orrorin* is still debated, however, because the scrappy fossils do not yet offer clear-cut, convincing evidence.

Ardipithecus

Originally placed in *Australopithecus*, *Ardipithecus* is a genus represented by an accumulating number of specimens from sites like Aramis in the Middle Awash region of Ethiopia. *Ardipithecus ramidus* means "root ape" in both Latin and in the native language of the region. There is little morphological change from 5.8 to 4.4 Mya, so those who work on these fossils (e.g., Tim White, Gen Suwa, Berhane Asfaw, Johannes Haile-Selassie) kept variation to the subspecies level with *Ardipithecus ramidus ramidus* (4.4 Mya) and *Ardipithecus ramidus kadabba* (5.8–5.2 Mya).

Little of the skull and postcranial morphology of *Ardipithecus* is preserved. There is a fragmentary skeleton waiting to be published, but for now there are mostly just bits of skull, teeth, jaws, and a few postcranial remains on record. The reduced size of the upper and lower canines relative to premolars and molars is more human- than ape-like. However, *Ardipithecus* has primitively thin enamel and primitive, albeit diminished, morphology of the CP3 honing complex.

Was *Ardipithecus* bipedal? The foramen magnum is located underneath the skull in *A. r. ramidus,* suggesting it was. Also there is a single toe bone (phalanx) of *A. r. kadabba* that is shaped like a human's as opposed to a chimpanzee's.

Evidence from *Ardipithecus* sites—like seeds, monkeys, and browsing antelopes—indicate the habitats were mostly closed canopy woodlands, not savannahs.

AUSTRALOPITHS

Members of the genus *Australopithecus* (which are often referred to as "australopithecines" or "australopiths") are undoubtedly bipedal and some of the species are considered direct ancestors to humans. Like the earliest hominins, australopiths and their descendents *Paranthropus* are only found in Africa (Figure 3.8). Thanks to the enormous fossil record of australopiths, with thousands of specimens including nearly complete skulls and skeletons, much is known about the genus.

Australopiths all share general features of the skeleton. They are smaller-brained than modern humans, at about half the size similar to chimps and gorillas. Adults were about 3–3.5 feet tall. They resembled upright-walking chimpanzees and may or may not have manufactured stone tools, but like living chimpanzees and gorillas they probably frequently used tools.

Australopiths are mainly known from cranial and dental fossils. The bones of the skull behind the eyes show very strong constriction ("postorbital constriction"), which makes room for large chewing muscles (the *temporalis* muscles) that attach from the jaw to the top of the head. Humans have very small chewing muscles and no postorbital constriction. Australopith faces are prognathic and the greatest breadth of the skull is at the base, toward the neck, unlike humans whose round skulls are widest up high (Figure 3.7).

In general australopiths have small incisors and canines relative to body weight in comparison to living apes. In this way they are more humanlike, but they still have larger teeth compared to humans. They

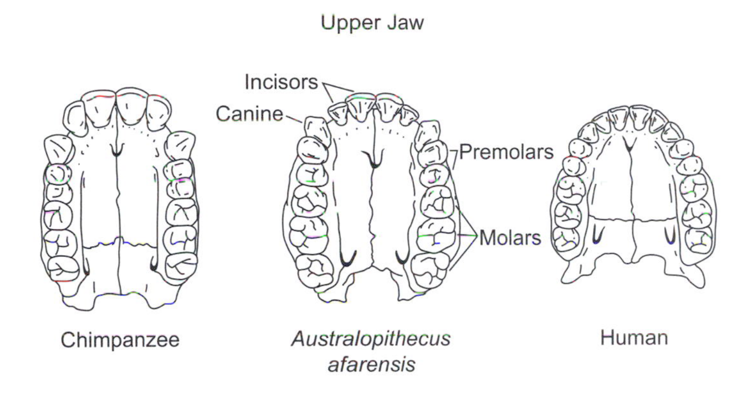

Figure 3.8 The human (*H. sapiens*) maxilla (upper jaw) shows the dental characteristics that distinguish *H. sapiens* from chimpanzees and australopiths including small vertical incisors, small flat canines, no diastema (gap) by the canine, flat molars with low round cusps, and curved tooth rows. The chimpanzee's (*Pan troglodytes*) upper jaw is similar to what we expect the teeth and jaws of the earliest hominins to resemble. Notice the large, spatula-shaped incisors, large pointed canines, straight parallel tooth rows, and large molars. *A. afarensis* shows traits that are intermediate between chimpanzees and humans. *Illustration by Jeff Dixon.*

have thick-enameled molars with bulbous cusps anchored in their thick, bony mandibles. Australopith body size is in the range of modern chimpanzees with a high degree of sexual dimorphism. They had human-like grip capabilities evident in their hand bones. It is debated whether or not australopiths were still adept tree-climbers or if they were obligatory bipeds because they retained some arboreal features like a funnel-shaped thorax, long curved hand and foot phalanges, relatively short legs, relatively long arms. But they also have evidence for bipedality like the distinct foot strike pattern (as evident by the Laetoli footprints in Tanzania), a broad flared ilium, a carrying angle to the femur, a foramen magnum tucked under the skull, a less thumb-like big toe that is more in line with the rest of the toes.

There are four species of *Australopithecus* that are generally agreed upon by most researchers. The genus *Australopithecus* arose in the Pliocene as early as 4.2 Mya with *Australopithecus anamensis* and lasts until about 2.5 Mya with *Australopithecus garhi*, a species that overlaps with the origin of our own genus *Homo.* Three of the australopith species are found in East Africa and one, *A. africanus,* is found at cave sites in South Africa.

It is important to note that the caves in South Africa that hold australopiths, *Paranthropus,* and some specimens of early *Homo* contain no evidence that the hominins were occupying those caves. Instead it appears as though their bones accumulated, just like the bones of other animals, by falling through the crevices and gaps of the ground into underground limestone caves. Carnivores, like leopards, and large birds of prey are blamed for many of these bone accumulations because they tend to use trees that grow near their openings. These are the same kinds of caves, formed by ground water dissolving the bedrock, that are common in America and that can eventually form sinkholes. Although nonhuman primates, like baboons, have been observed seeking shelter in caves, real evidence of human cave occupation does not occur until the middle Pleistocene with Archaic humans.

Australopithecus anamensis

The oldest australopith is *A. anamensis.* Fossils of *A. anamensis* date between 4.2 and 3.8 Mya and come from Kanapoi and Allia Bay, Kenya, and Asa Issie, Ethiopia. Today the Kenyan sites are located near the shore of the southern tip of the large Lake Turkana. At various times in the past, parts of the lake were greatly reduced and these sites were located in what were once deltas and floodplains of the ever-changing landscape. When the first fossils of *A. anamensis* were discovered by Meave Leakey, Kamoya Kimeu, and Alan Walker in the mid-1990s, they were the earliest known bipedal hominin fossils.

The mandible from Allia Bay is u-shaped like an ape. It has a large bony buttress at the mandibular symphysis (i.e., the joining of the two halves of the jaw, right under the incisors) to protect it from breaking like a wishbone during chewing. It has larger teeth than chimpanzees and later *A. afarensis.*

A. anamensis shows gorilla-like sexual dimorphism in canine size and body size. There is a single fragmentary humerus from Kanapoi that lacks the knuckle-walking morphology of African great apes. The radius is longer than even the longest chimp radius. The tibia is oriented at the ankle like modern humans, strongly indicating bipedal locomotion was used. Further evidence that *A. anamensis* fossils do not belong to the chimpanzee lineage comes from the skull near the ear where the hole for the chorda tympani (the facial nerve for taste) passes through the bone differently than it does in chimpanzees.

The teeth and bones of *A. anamensis* are very similar but look primitive in comparison to later *A. afarensis.* Therefore, *A. anamensis* is widely considered to be ancestral to *A. afarensis.*

Australopithecus afarensis

The australopith with the richest fossil record is *A. afarensis* and the best-known specimen is a partial skeleton affectionately called "Lucy" (AL 288-1). She was discovered by Donald Johanson in 1974 at the site of Hadar in the Afar region of Ethiopia and she dates to about 3.2 Mya. An estimated 40 percent of her skeleton is preserved if her missing hand and foot bones are not included. In another spectacular find at Hadar, several individuals were discovered that died together 3.2 Mya and this group of *A. afarensis* is called The First Family because it confirmed that the species lived in social groups, but the cause of death is still unclear.

A. afarensis lived from approximately 4 to 2.5 Mya in an area stretching from Ethiopia to Tanzania to Chad. *A. afarensis* fossils are found at sites that were once predominantly open and dry savannahs and woodlands. Cranial capacity ranges from 400 to 500 cc and adult body masses average 29 kg for females and 45 kg for males (about 60 and100 lbs respectively). Males had nearly twice the body size of females much like modern gorillas. They had long arms, short thumbs, and curved fingers and toes.

The teeth have some ape-like characteristics, including large pointed canines, and a small diastema in the lower jaw to accommodate the upper canines. The molar cusp patterns are intermediate in character between apes and humans and the cusps themselves are low and round like humans. Overall the teeth are smaller than in apes and the tooth row is not as parallel as in apes or *A. anamensis* (Figure 3.9).

The postcranium of *A. afarensis* is characterized by ape-like traits like short legs, long arms, and a funnel-shaped chest. But it also had human-like adaptations for walking upright like a more bowl-shaped pelvis and less curvature in the fingers and toes than apes (although still curved). This mixture of ape-like and humanlike features is what make *A. afarensis* the classic link between apes and humans—a true bipedal ape. However, exactly how arboreal this species remained is still debated.

Evidence for bipedalism is known from many more anatomical regions than in previous species, partly because of better preservation and sheer numbers on record, but also because bipedalism is probably more

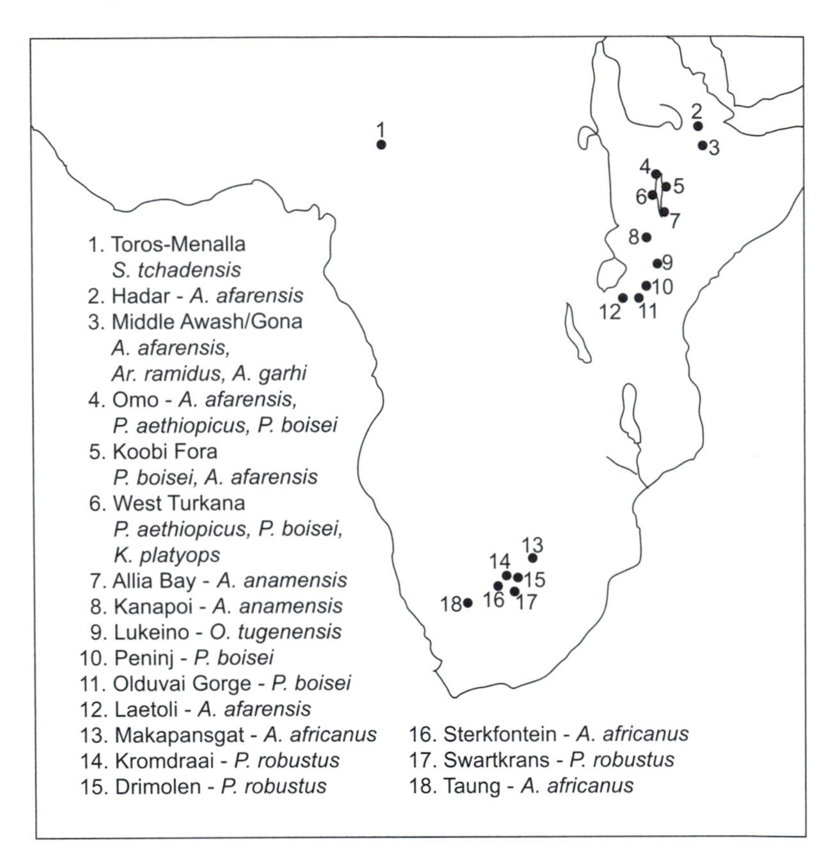

Figure 3.9 A map of sub-Saharan Africa shows sites where early hominins have been found. *Illustration by Jeff Dixon.*

evolved in *A. afarensis*. There is a well-preserved knee joint (the end of the femur and the matching top of the tibia) from a specimen at Hadar, the same site that produced Lucy, which clearly shows the angled knee of a biped (Figure 3.6).

Footprints attributed to *A. afarensis* also show evidence of advanced bipedalism. At the site of Laetoli, Tanzania, a trail of hominin footprints is preserved in a hard ash layer which is dated to about 3.7 Mya, within the time and geographic span of *A. afarensis*. Mary Leakey led the excavation of the footprints of an array of animals including rabbits and elephants. The hominin ones, which cross the path of a three-toed horse, resemble a modern human's. There is a depression for a strong heel strike just like a modern bipedal stride and there is also the appearance of an arch, something unique to the human foot that stores energy like a spring. However, the big toe is separated slightly from the rest of the toes.

Kenyanthropus Platyops

Like the discovery of *Sahelanthropus*, new skulls are sometimes the basis for the christening of a new hominin taxon. In the case of *Kenyanthropus platyops*, a new genus was named for a distorted and highly fragmented skull (KNM-WT 40000). Meave Leakey led the team of scientists who interpreted the 3.5-million-year-old skull from near Lake Turkana to mark the beginning of a diet-driven adaptive radiation—a second lineage beginning in the middle Pliocene and including the early Pleistocene specimen of *Homo rudolfensis* (KNM-ER 1470; Figure 3.11, right) with which it shares its characteristically flat face. This flat face, and hence the meaning of the name "platyops," is associated with a different diet from *A. afarensis*—the only other hominin present in East Africa at 3.5 Mya. Critics argue against using the single, highly cracked and fragmented *K. platyops* skull to establish an entirely new lineage of hominins during australopith times. Currently this taxon is in a sort of paleoanthropological probationary period while it awaits support from future fossil finds and analyses.

Australopithecus africanus

The name of this hominin literally means the "southern ape person of Africa" and is found at cave sites like Swartkrans, Sterkfontein, and Makapansgat in South Africa. *A. africanus* first appeared around 3 Mya and lived until about 2.5 Mya years ago in regions of open woodlands and grasslands. These hominins are similar to *A. afarensis* in the size and shape of their skeleton, and the difference in body size between males and females.

The skulls, however, have rounder jaws, smaller canines, flatter faces (less prognathism), and none of the bony crests seen in *A. afarensis* or *Paranthropus*. Their faces display characteristic pillars on either side of the nose. In addition, their brains were larger—averaging about 500 cc. Because of their similarities with later hominins, *A. africanus* is often considered ancestral to early *Homo*.

The most famous *A. africanus* fossil is the Taung child, a skull and partial endocast. Discovered in South Africa in 1924, this fossil was initially rejected as a human ancestor because the scientific community at the time assumed that large brains were the first trait to evolve in humans. Despite the small brain, the teeth of the Taung child are

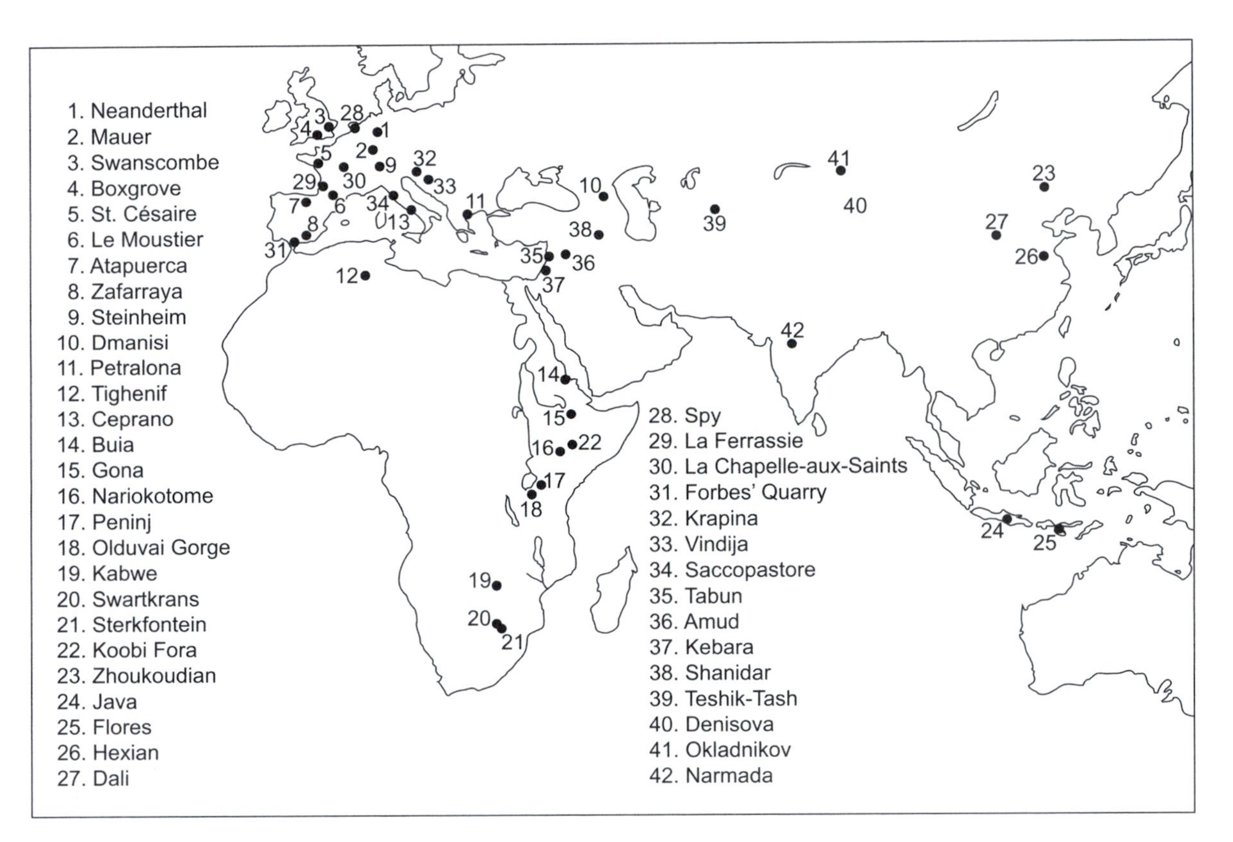

Figure 3.10 A map of the Old World shows sites where fossils of the genus *Homo* have been discovered. Not included are fossils sites bearing modern humans since those became widespread after 25 Kya. *Illustration by Jeff Dixon.*

humanlike. Plus, the location of the foramen magnum shows that the skull sat directly on top of the body and suggests that the individual walked upright. The authenticity of the Taung find was not widely accepted until the 1940s when adult fossils with the same mix of characteristics were discovered.

Further inspection of the Taung Child shows interesting features that are consistent with other australopiths. The frontal bone (forehead) recedes less than an ape, the deciduous ("milk" or "baby") canine is smaller like a human. Based on the schedule of tooth eruption it is estimated to have been about three or four years old. The body size, if extrapolated to an adult, would have been like a chimpanzee. The endocast was measured to be about 400 cc which, when grown-up to adult-size, would be about 450 cc.

The postcranial skeleton of *A. africanus* shows clear adaptations to bipedalism. Sts 14, found in 1947, is a partial skeleton with the vertebral column, the pelvis, and part of a femur from Sterkfontein, dating to about 2.5 Mya. The vertebrae show clear adaptations for bipedalism and it has a humanlike bowl-shaped pelvis.

Recently Ron Clarke discovered an astonishingly complete skeleton in Sterkfontein cave in South Africa that dates to *A. africanus* times. The individual came to be known as "Littlefoot" (STW 573) because its foot was discovered early on before the rest of the skeleton was dug out of the rock. It is still being meticulously cleaned with airscribes and chemicals. However it is clear that Littlefoot will eventually be revealed as the most complete specimen of an early hominin on record.

Australopithecus garhi

Named "surprise" in the local Afar language for its intriguing combination of traits, *A. garhi* is known from a single site in Ethiopia called Bouri in the Middle Awash. It dates to 2.5 Mya and was announced by Berhane Asfaw, Gen Suwa, and Tim White in 1999. The *A. garhi* bones are a significant find because at this time period in the East African hominin fossil record, there are very few nonrobust australopiths, and also this is about the time early *Homo* was just starting to evolve there.

A. garhi is known from a skull and some postcranial bones. The skull has a sagittal crest and a cranial capacity of 450 cc. The maxilla is surprisingly large and robust. It has primitive frontal, facial, and subnasal (below the nose) morphology relative to *A. africanus* and early *Homo*.

The teeth are like giant early *Homo* teeth. The incisors are more similar in size to the molars compared to earlier australopiths—a condition which is also more humanlike.

The limb bones indicate *A. garhi* had small stature like earlier australopiths. However they were more muscular since the muscle attachment sites on the bones are larger. The limb proportions show that *A. garhi* had relatively long forearms like earlier australopiths, yet the femur was much longer, a bipedal adaptation shared by *Homo*. The Bouri remains document the earliest-known increase in femur length, which points to clear selection for more efficient bipedality. The proportions are a mosaic of ape and human morphology since the ratio of humerus to radius/ulna is ape-like, but the ratio of humerus to femur is humanlike.

Faunal remains from Bouri have stone tool cut-marks but there are no stone tools from the site. Not only is *A. garhi* the first hominin to have longer legs but it is also the first to be associated with the use of tools to procure highly nutritious marrow from animal bones.

PARANTHROPUS (ROBUST AUSTRALOPITHS)

It is becoming more popular to refer to the robust species of the australopiths with their own genus *Paranthropus*. According to the rules of Linnaean classification, groups must share a single ancestor, but it is possible that robust australopiths evolved independently in East and South Africa from the australopiths in each region (*A. afarensis* and *A. africanus* respectively), meaning they could have separate roots. But for clarity and for continuity with current trends, here the robust australopiths are referred to as *Paranthropus*.

Although morphological links between the australopiths and *Paranthropus* are evident, there is no mistaking *Paranthropus* with their distinct skull and tooth morphology. *Paranthropus*, which includes *Paranthropus aethiopicus*, *Paranthropus boisei*, and *Paranthropus robustus*, makes up an evolutionary dead end in the hominin phylogeny. It is an extinct lineage that branched off from australopiths around 2.5 Mya and then hung around long enough to coexist with early *Homo* in East and South Africa until about 1 Mya.

Paranthropus cranial capacities range from 410 to 530 cc. Although they shared an enlarged brain size and bipedal capability with *A. africanus* and *A. afarensis*, species of *Paranthropus* had markedly larger teeth and jaws.

Paranthropus used their large flat molars in their large jaws to process fibrous foods. Four of a modern human's molars would fit on one

Paranthropus molar, yet *Paranthropus* was only half as tall as average adult male humans. Their skulls boast prominent attachments for the chewing muscles (*temporalis*), including a sagittal crest along the top of the skull (similar to a small bony mohawk). Their large, flared cheekbones and postorbital constriction make room to accommodate those same muscles running from the jaw to the sagittal crest, which would have been as big as a pound of steak on either side of the head. Based on the cranial and dental morphology and also on isotopic analysis of the composition of tooth enamel (see Chapter 5), we know that *Paranthropus* ate a diet of tough fibrous foods and hard seeds, nuts, and fruits. Average adult body masses were about 33 kg for females and 45 kg for males (range of about 70–110 lbs) with a stature comparable to that of *A. afarensis.*

P. boisei (pronounced boy-zee-eye) is probably the most well known of the genus as it boasts the complete cranium (OH 5) that Louis Leakey found at Olduvai Gorge in 1959. (OH stands for "Olduvai Hominid".) Leakey originally placed OH 5 in a new genus *Zinjanthropus* and it earned the nickname "Zinj." The OH 5 skull was the first hominin to be accurately dated (1.75 Mya) by the potassium/argon method. Out of the three *Paranthropus* species, *P. boisei* has perhaps the most exaggerated molars compared to the size of the small incisors, and has "hyper-robust" cranial crests for the chewing muscles, which led to the nickname "Nutcracker Man" (Figure 3.7). The species ranged from 2.5 to 1 Mya and specimens of *P. boisei* are also found in Tanzania and at Koobi Fora, Kenya.

P. robustus fossils come from the South African sites of Sterkfontein, Swartkrans, and Drimolen and have skulls and teeth that are like less exaggerated versions of *P. boisei.* They spanned from about 2.0–1.5 Mya. *P. robustus* is affiliated with a record of bone tools at the Swartkrans cave site. In the first half of the 20th century, Raymond Dart interpreted these tools as weapons that were part of an "osteodontokeratic" (bone-tooth-horn) culture used by "killer" ape-men. However, since Dart's time, much more observations of chimpanzee behavior have been made and new microimaging techniques for artifacts and bones have become available. With modern knowledge it is clear that the bone tools from Swartkrans were used for digging tubers out of the ground and some were also used as wands for fishing termites out of their mounds, similar to the way chimpanzees obtain the insects with twigs.

The beautifully complete skull that Alan Walker discovered in the early 1980s on the west side of Lake Turkana, Kenya, is the best known specimen of the earliest species, *P. aethiopicus* (2.7–2.3 Mya). It is known as the "Black Skull" (KNM-WT 17000) for the color it became as it fossilized. (KNM stands for Kenya National Museum and WT stands for West Turkana.) The cranium resembles an aerodynamic, road-hugging sports car because of its prognathic face (like the hood) and its sharp cranial crests, particularly the horizontal one across the back of the head (like the spoiler). *P. aethiopicus* brain size is estimated to have been very small at about 410 cc and it is considered to be ancestral to *P. boisei* and *P. robustus*.

There is no clear explanation for why *Paranthropus* went extinct about 1 Mya. It is intriguing that a group of small bipedal vegetarians would go extinct just as taller, carnivorous *H. erectus* was spreading across the Old World. What a striking parallel to the present-day vanishing of great apes in the presence of humans.

THE HUMAN GENUS

With the genus *Homo* we are given the earliest fossil evidence for dispersal outside of Africa (Figure 3.10). As with all lineage beginnings, the identification of the first member of the genus *Homo* in the fossil record is difficult. There are several specimens from Ethiopia, Kenya, and Tanzania that are worthy candidates. The problem lies in the definition of *Homo*. The characteristics that warrant inclusion in the exclusive club of the human genus are not generally agreed upon.

It is no longer fashionable to assign hominins to *Homo* based on stone toolmaking ability, since it has been difficult to conclusively demonstrate association between the earliest tools and a specific species. There is considerable overlap in geography and age of *Homo*, *Paranthropus*, and *Australopithecus* during the late Pliocene and early Pleistocene, so we cannot definitively exclude *Australopithecus* or *Paranthropus* from the stone toolmakers' guild.

Other researchers rely on anatomical rather than behavioral evidence for inclusion in the genus *Homo*. The majority of researchers consider *Homo habilis* to be the earliest member of the human genus based on its differences from *Australopithecus*, such as its smaller and flatter face, less sloping forehead, more rounded skull, smaller jaws and teeth, thinner cranial bones, and lack of cranial crests. The long-standing brain size requirement is still generally followed: skulls that have cranial capacities over 600 cc are safely considered to belong in the genus *Homo*. As long as skulls with that cranial capacity do not have *Australopithecus* or

Paranthropus features, that rule is supported and none have been found to contradict it so far.

Based on its more ape-like limb proportions reconstructed from fragmentary skeletal remains, and based on many of its similarities to australopiths as opposed to later *H. erectus*, *H. habilis* is increasingly being placed in the genus *Australopithecus*.

One reason behind this taxonomic shift is the partial skeleton OH 62, a *H. habilis* from Olduvai Gorge. The fragmented remains of this old female were found near a dirt road after they had been unintentionally driven over. When the limbs are reconstructed and proportions are estimated, OH 62's body is very ape-like, sometimes even more ape-like than some australopiths. Some fossils, like this one in particular, cause more trouble than provide answers. Until better skeletons of *Australopithecus*, *Paranthropus*, and early *Homo* are discovered, the classifications of many fossil hominins around this crucial period of human evolution around 2 Mya are best considered tentative. In the end *Homo* is just a name. We know that at some point *Homo* evolved from earlier australopiths. The more fossils we find, the blurrier the division between "apeman" (australopiths) and "human" (*Homo*).

Fossil hominins clearly located within the early part of the genus *Homo* are also undergoing identity crises. *H. habilis* was established in 1964 by Louis Leakey, Philip Tobias, and John Napier. They designated the new species based on hand bones and skull fragments of a juvenile (OH 7) that they excavated from Bed I at Olduvai Gorge. *H. habilis* literally means "handy man"—a name that is based upon the inferred humanlike grip capabilities of the fossil hand. The species was deemed responsible for the Early Stone Age tools (Oldowan Industry) found near the site.

H. habilis is a species of small-brained, small-bodied hominins that is usually kept separate from another species that spanned the same time range (2.3–1.8 Mya) but had a larger brain, larger teeth, and a flatter face, *H. rudolfensis*. The two fossils that symbolize the differences between *H. habilis* and *H. rudolfensis* are the very small skull KNM-ER 1813 (500 cc) and the significantly larger skull KNM-ER 1470 (775 cc) (Figure 3.11). For some paleoanthropologists (splitters), the size difference between the two is too large to keep them in one species, but others (lumpers) consider it normal variation worthy of a single variable species. In 2006, new dates for the sediments at Koobi Fora were reported by Patrick Gathogo and Frank Brown. Now the smaller skull (1813) is deemed 0.25 Mya younger (1.65 Mya) than the larger one (1.9 Mya). If these dates are confirmed, there may no longer be an issue.

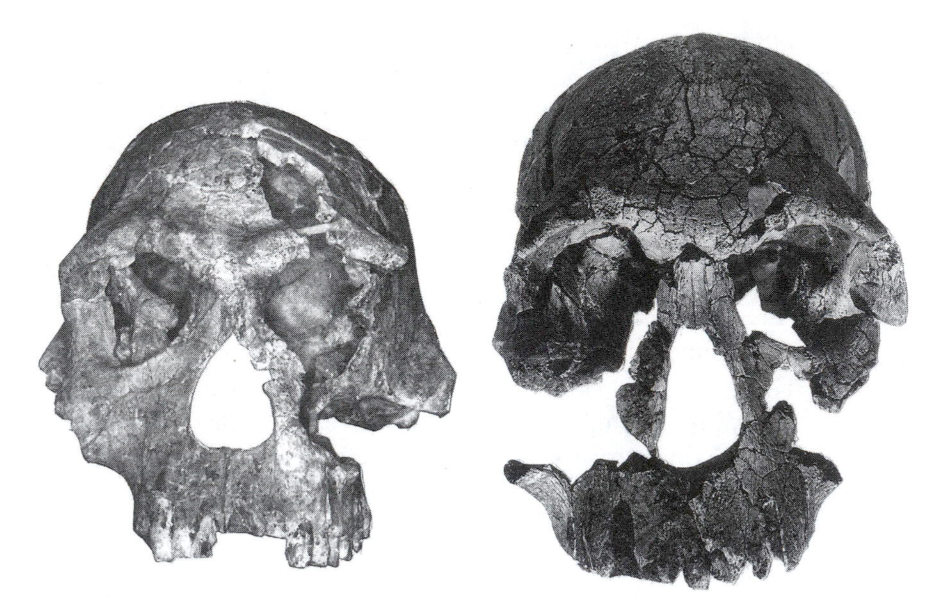

Figure 3.11 Both found in the early 1970s at contemporaneous sites at Koobi Fora, Kenya, these skulls (right: KNM-ER 1470, left: KNM-ER 1813) typify the diversity of size and shape in early *Homo*. *Photographs by Alan Walker.*

HOMO ERECTUS

The first *H. erectus* fossils to be discovered were a skullcap (the top portion of the cranium that does not include the face or the bottom where the foramen magnum is located), a molar, and a femur from Trinil, Java, in 1891. These finds comprise the missing link that Eugène Dubois set out from the Netherlands to find. Since then, paleoanthropologists have collected *H. erectus* specimens from Ethiopia, Kenya, Tanzania, South Africa, Morocco, Italy, India, China ("Peking Man"), and Indonesia ("Java Man" and "Solo Man"). The species marks the first dispersal of hominins outside of Africa that corresponds to a major change in ecology (see the section on "Scavenging and Hunting" in Chapter 5).

It is widely assumed that *H. erectus* evolved in Africa because the hominins that preceded it are found there. However, the earliest dates for *H. erectus* from sites in the Republic of Georgia and Indonesia are pushing 1.8 Mya. In order for the African origin of the species to be overturned, older fossils of *H. habilis* or australopiths would need to be found outside of Africa and, as of now, they have not.

H. erectus spans nearly 2 million years of hominin evolution, with the last bastion of the species holding on until about 30,000 years in Indonesia. *H. erectus* is distinguished from preceding hominins by several characteristics. The cranial capacity is increased with a range from about 750 cc to 1200 cc, which extends to within the range of modern human brain sizes (Figure 3.7). *H. erectus* skulls display no chins, receding foreheads, and massive, short projecting faces surmounted by well-developed browridges (which actually look more like a single horizontal visor compared to later hominin browridges which are smaller and sometimes 'm' shaped). They also had smaller teeth and thicker cranial bones than their predecessors.

With the arrival of *H. erectus* comes the first appearance of humanlike body size and proportions. The skeleton was nearly indistinguishable from modern humans, although it was more robust than ours. Males and females were more equal in body and tooth size and shape.

The most complete skeleton on record of a *H. erectus* is the Nariokotome boy, which is also sometimes called the Turkana boy (KMN-WT 15000) (Figure 3.12). Both names refer to the geographic region where the skeleton was excavated—the site of Nariokotome on the west side of Lake Turkana, Kenya. In 1985, Kamoya Kimeu discovered the boy by spotting a small chunk of the skull that led the excavation team to uncover the entire skull just near the surface and then to the rest of the skeleton deeper in the ground.

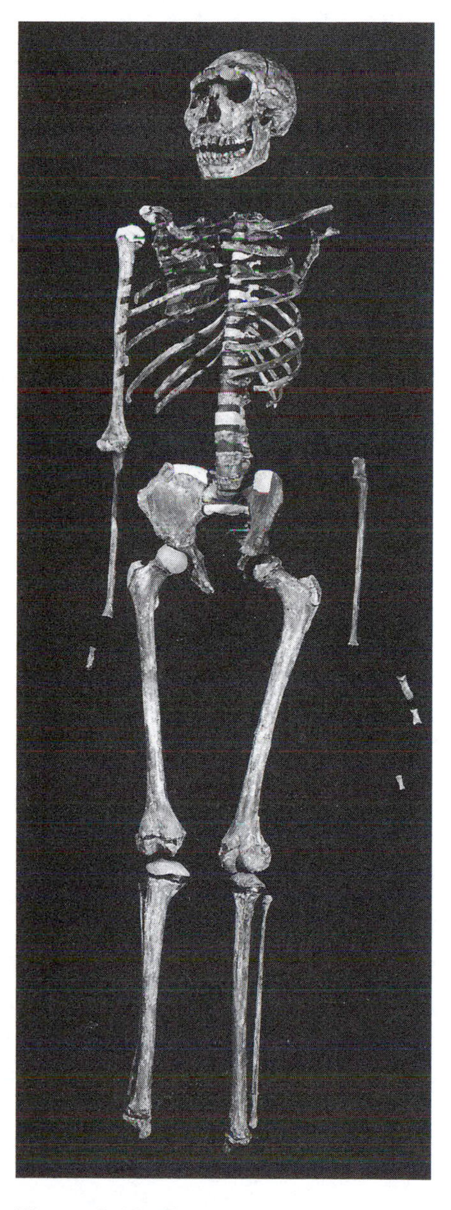

Figure 3.12 The nearly complete skeleton of a young *H. erectus* known as the "Nariokotome boy," and also the "Turkana boy," was discovered by Kamoya Kimeu on the west side of Lake Turkana, Kenya, in the early 1980s. *Photograph by Alan Walker.*

The microscopic structure of the Nariokotome boy's teeth indicate he died when he was about eight years old, not eleven or twelve as previously thought. At eight years old, the Nariokotome Boy stood 5 feet 3 inches and would have grown to be over 6 feet tall had he lived to adulthood. The large pitted scar on his mandible from a tooth root abscess could very well indicate the cause of his death. Lesions like that, if left untreated, can lead to infection which can spread to the bloodstream and poison the body through a life-threatening condition called septicemia.

The skeleton of the Nariokotome boy shows adaptations for fully committed and efficient bipedalism with no lingering arboreal traits. His limb proportions indicate that Allen's rule (see Chapter 5) was at work on humans way back in the early Pleistocene; his long distal limb segments and narrow pelvis are similar to those of modern people that live in hot tropical climates.

The Asian *H. erectus* skulls share common elements of their appearance. They have even more pronounced browridges, more pronounced keels all over the skull, and thicker cranial bones. This variation causes some researchers to lump all *H. erectus* from all over the Old World into one species spanning from 1.8 Mya to 30 Kya. But those who emphasize the differences in the Indonesian fossils keep those in *H. erectus* but call the fossils from Africa and Georgia *Homo ergaster*, which means "working man."

DMANISI

Located in the foothills of the Caucasus Mountains, the village of Dmanisi, Georgia initially interested archaeologists because of its 9th century medieval fortress. In the early 1980s, fossils were discovered in the walls of the storage pits that the medieval inhabitants had dug within the fortress (Figure 3.13). One of the fossils was identified as a *H. erectus* jaw. Since then at least four skulls and one partial skeleton of *H. erectus* as well as hundreds of crude stone tools have been unearthed at Dmanisi (Figure 3.14).

The fossil-bearing layer at Dmanisi is dated to about 1.78 Mya, tying it with Trinil, Indonesia, for the earliest hominin site outside of Africa. Because of the consistent flow of hominin fossils from the site, Dmanisi is the premier locality for studying the first hominin dispersal out of Africa.

From the few fossils recovered so far, it is clear that there were big and little individuals. Considering the variation displayed by earliest *Homo* (*habilis* and *rudolfensis*), which overlaps in time, perhaps such variation was normal sexual dimorphism or normal population variation. Perhaps there

Figure 3.13 Fossil deposits underneath a 9th-century fortress at Dmanisi were discovered while archaeologists explored medieval storage pits like this one. *Photograph by Holly Dunsworth.*

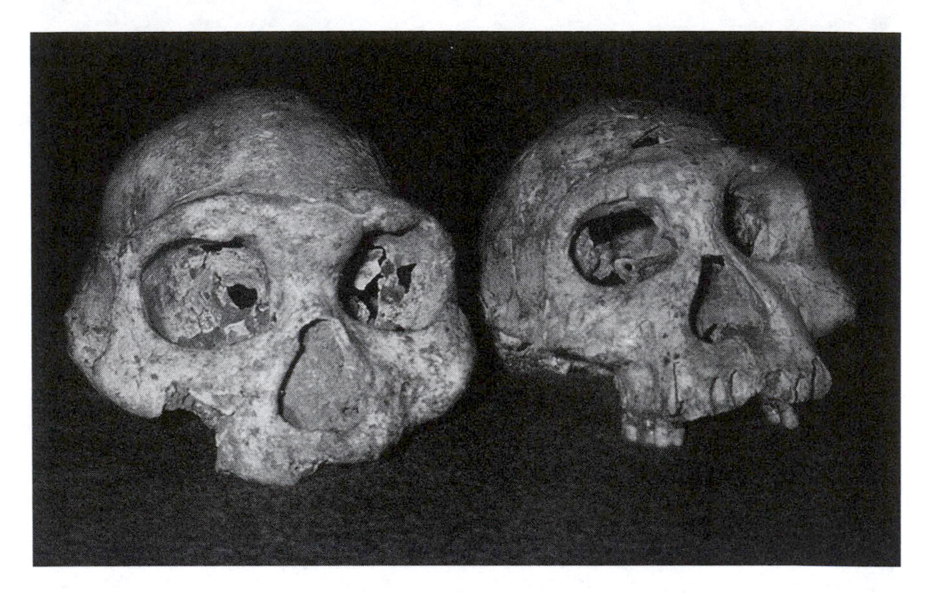

Figure 3.14 Two of the skulls discovered at the early Pleistocene site of Dmanisi, Georgia, are excellently preserved (D 3444 (left) and D 2700 (right) affectionately known as "Poor Marc"). *Photograph by Holly Dunsworth with permission from David Lordkipanidze.*

are two species of hominin at Dmanisi. The stone tools at Dmanisi are very crude Oldowan lithics. They are not the Acheulean tools that are normally associated with *H. erectus.*

ARCHAIC *HOMO SAPIENS*

Hominins that lived all over the Old World between 800 and 125 Kya belong to a category called "Archaic *H. sapiens.*" Archaics are a transitional group between *H. erectus* and modern humans and include the Neanderthals, which are discussed separately.

Archaics were still anatomically distinct from modern people, mostly in their skulls, which were thick-boned and low-vaulted, and featured prominent browridges, sloping foreheads, and small chins. Reminiscent of earlier *H. erectus,* their skulls retain their long and low profile, but show increased roundness like modern humans with the maximum breadth higher than the ear holes. Sporting balls are good metaphors for the shape contrasts between the species' crania (the skull without the face). *H. erectus* crania are like American footballs, Archaics' are like rugby balls, and modern humans' are like round soccer balls.

The Archaic face is still prognathic but much less so than earlier hominins. Their teeth were somewhat larger than those of modern humans, although they were markedly smaller than the teeth of earlier hominins. Archaic skeletons are robust due to heavy musculature in life, and their brain size averaged about 1,200 cc (ranging from 900 to 1400 cc), which is practically the same as modern human brain sizes. Despite physical similarities to modern *H. sapiens*, they lacked the cultural capacities that distinguish our species. Archaic fossils are found in association with Acheulean and more complex tools but not the most complex ones that modern humans invented.

It is not certain which Archaics led to Neanderthals or modern humans but despite the ambiguity of the overall category, there are a few fossil groups which are considered their own species. *Homo heidelbergensis* includes the majority of Archaics in Europe, the Mediterranean, and Africa dating between 600 and 125 Kya. The name *H. heidelbergensis* was given to a mandible found in Heidelberg, Germany, in 1907. But, the earliest members of the species include African skulls found at Kabwe in Zambia and Bodo in Ethiopia, which dates to 600 Kya and is associated with Acheulean artifacts. Archaic fossils from Asia come from sites like Dali in China dated to about 200 Kya. Asian Archaic skulls share some primitive traits with *H. erectus* but have larger cranial capacities and are not prognathic.

Homo heidelbergensis in Europe probably gave rise to Neanderthals. A site in Atapuerca, Spain, dating to about 800 Kya has produced several skeletons of *Homo antecessor* that are potential, very early precursors to Neanderthals. Numerous skeletons were also discovered at the nearby site of Sima de los Huesos ("pit of bones") dating to a later age of 350 Kya and these may or may not be descendents of the earlier populations or ancestors of Neanderthals. Earlier in the fossil record, a skull from Petralona, Greece, dated to about 400 Kya, could be ancestral to Neanderthals with its huge browridges and large nose opening. An even older skull from Arago, France, dated to about 450 Kya, is also Neanderthal-like with its large, relatively projecting face.

Archaics are the first hominins to have associated archaeological evidence of constructed shelters either from old post-holes or preserved materials. In Lazaret Cave, France, a shelter was built against a cave wall and in Terra Amata (also in France) there is evidence of a seasonal or short-term shelter use. With Archaics, we find the first clear evidence of hominin cave use.

At this point, evidence for big game hunting is also abundant. At 250 Kya at the site of La Cotte de St. Brelade in the Channel Islands between

Great Britain and France, there are skeletons of mammoths and wooly rhinoceroses, associated with numerous stone flakes. The carcasses were processed for eating, the skulls were cut open to extract the brains, and bones were burnt. The numerous skeletons in the collection suggest they were driven off cliffs and since there is no evidence of long-term hearths, the site was probably a temporary hunting camp.

The earliest preserved spears also appear at this time, around 400 Kya, and come from Schöningen, Germany. They were crafted from the inner part of the trunk of a spruce tree and have fire-hardened tips and tapered handles. The balance of the spears suggests they were weighted for throwing, but it is just as possible they were only thrusted during use. The spear site also contained hundreds of horse bones that had been killed (probably with the spears) and then butchered.

NEANDERTHALS

The first nonhuman hominin fossils known to science were Neanderthals (pronounced Nee-an-der-talls) or *Homo neanderthalensis.* They were a stocky, muscular, barrel-chested species known from sites in Europe, the Middle East, and Eurasia. The name Neanderthal comes from the location of one of the earliest discoveries of the species, in the Neander Valley, Germany, in 1856. Based on misinterpretations of the anatomy and the desire to elevate humans from our evolutionary cousins, Neanderthals were originally regarded as ape-like savages with little intellect and a stooping, knuckle-dragging walk. The stereotypes have been carried over today, despite all of the modern evidence to the contrary.

Neanderthals existed from about 350 to 28 Kya and many of the sites discussed under the Archaics are also considered Neanderthal sites. Toward the end of their existence Neanderthals became increasingly distinct in appearance from modern humans, increasingly similar in culture to modern humans (this point is arguable), and then they disappeared. The "classic" Neanderthals, the most anatomically distinct and stereotypical specimens, are found in Europe. The skull has unmistakably large, rounded, m-shaped browridges that loom over a large nose opening. The middle of the face projects at the nose and appears pulled-out compared to the rest of the face, which appears to sweep back from the nose at the cheekbones (Figure 3.7). The jaw has no bony projection for a chin like humans and also has an extra space called a retromolar gap behind the third molars. The molars are called "taurodont" because of their sturdy and puffed-up, or "bull-like," appearance. The low, flat frontal bone, long braincase, and bulging occipital bun (i.e., bump) holds the largest brain, on average—for all hominins including humans—at 1300–1600 cc.

The postcranial skeleton of Neanderthals is also unmistakable. The extremely robust bones indicate that Neanderthals were strong, muscular, and athletic. The long bones of Neanderthals are thick and curved because of the great muscular forces placed on them. Originally, these curved bones were blamed on rickets, a disease associated with poor nutrition.

Neanderthal skeletons often show evidence of trauma. Eric Trinkaus likened their particular pattern of injuries to the types of broken and healed bones experienced by American rodeo riders. Neanderthals were not necessarily riding wild animals, but they were probably coming into dangerously close contact with them. Studies of the arm bones by Steven Churchill even suggest that they preferred to thrust spears into their prey rather than throw them, which lends support to the interpretation of their dangerous hunting techniques. Neanderthal teeth worked just as hard as their bodies. Often their incisors are worn down and even show stone tool cut-marks across them as if Neanderthals used their jaws as a vice to prepare food or animal hides.

In spite of their rough lifestyles, Neanderthals had advancements in cultural materials that approached modern humans. One of the most recent (or last) Neanderthal sites at Grotte du Renne in France dated to 34 Kya has preserved perforated and grooved teeth, ivory, and bone beads that were probably used for personal adornment. If it is traditionally thought that only modern humans behaved like this, should it be assumed that Neanderthals were trading or interacting with modern humans? It is debated whether or not Neanderthals and modern humans evolved this behavior independently or if they ever interacted.

A circle of mammoth bones with hearths, tools, and animal bones at a site in the Ukraine show that Neanderthals made camps like modern humans. Therefore, the strictly "caveman" image of Neanderthals could be the result of the preservation bias afforded by caves. There are no needles yet associated with Neanderthals so it is possible they did not fashion clothing. However, that does not preclude the probability that they wore animal pelts since they lived in cold climates.

It is not known why Neanderthals disappeared around 28 Kya. Perhaps they became too well adapted to the cold so that they could no longer survive between Ice Ages. Perhaps modern humans were better at filling

Table 3.2 Major Stone Tool Types and Their Time Periods in the Paleolithic, or Old Stone Age

Oldowan	Lower Paleolithic	2.5 Mya–300 Kya
Acheulean		
Mousterian	Middle Paleolithic	300–40 Kya
Modern human toolkit	Upper Paleolithic	40–10 Kya

a similar niche and out-competed them. Or, perhaps modern humans swamped the Neanderthal gene pool by interbreeding with them.

A new hypothesis by Steven Kuhn and Mary Stiner suggests that the demise of Neanderthals could be blamed on their method of hunting. Or alternatively, the success of modern humans could be due, in part, to their implementation of a division of labor between the sexes. Evidence suggests that entire Neanderthal populations were involved in hunting big game. This is quite the opposite of early modern humans who—modeled after living hunter-gatherer or foraging societies like the San people of the Kalahari Desert of southern Africa—had the men hunt large game while the women gathered small game and plants. This division of labor ensures a continuous and diverse food supply since the obtainment of meat from large animals can be patchy and unpredictable. Neanderthals were hunting perilously with spears (no bows and arrows or even spear-throwers yet) and they were coming in close contact with their prey. It is possible that women and children were participating. There are no bone needles, very few small animal remains, and no grinding stones for preparing plant foods at Neanderthal sites. These are the artifacts associated with the female roles at home bases in human societies and would be left behind if women and children were staying back from the hunts. The Neanderthal strategy may have been aimed at nutritiously rich protein sources, but it could have been too specialized and could have also endangered their reproductive core to the point of species demise.

STONE TOOLS

Stone tools become smaller, more specialized, and more difficult to make through time (Figure 3.15; Table 3.2). The earliest, most primitive, stone tool industry recognized in the archaeological record is the Oldowan industry or the Early Stone Age tradition. It is a culture of simple crude cores and sharp flakes with only a handful of named types in the toolkit like choppers, hammerstones, and scrapers. Flakes held between the thumb and the forefinger can be used like a scalpel to skin

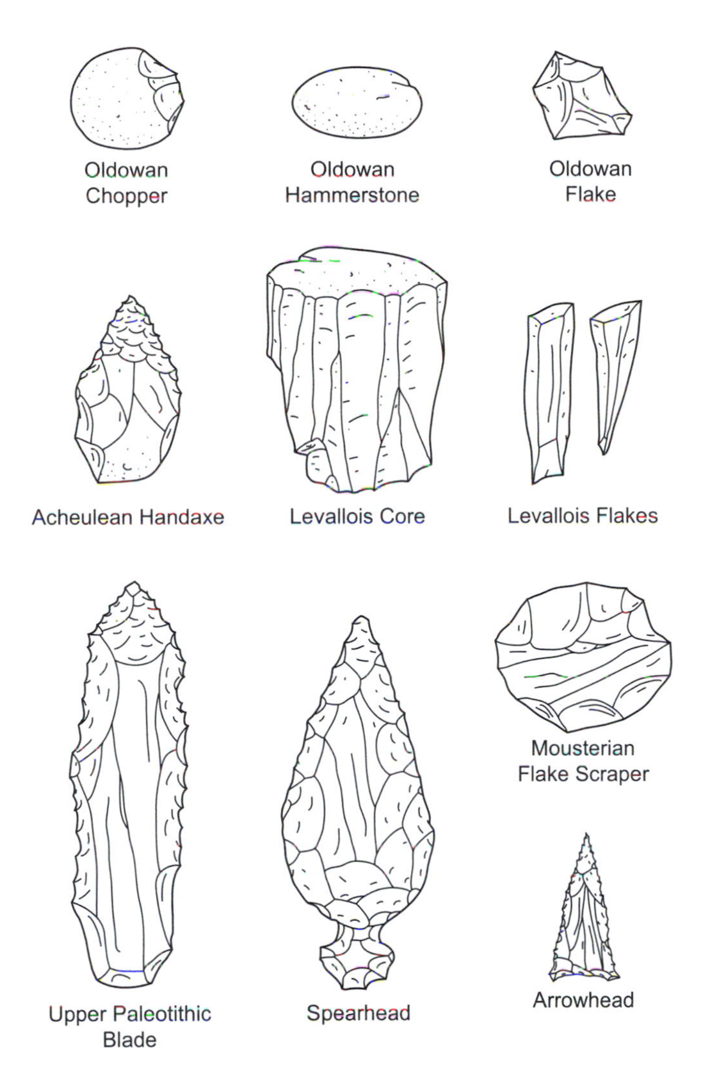

Figure 3.15 A comparison of stone tool types found throughout the prehistoric record. Spearheads and arrowheads would have been hafted onto wooden implements and were not invented until 50 Kya. *Illustration by Jeff Dixon.*

and butcher an animal carcass. Intentionally manufactured flakes stand apart from ones made by Mother Nature in their scars from production: because of the classic fracture pattern, there is usually a mark from where the hammerstone struck the flake from the core and underneath that is a bulge or a "bulb of percussion."

Oldowan scrapers and choppers are polyhedral and spherical implements that would have been useful for a wide variety of food preparation processes, of both animals and plants. Hammerstones not only knock flakes from softer rocks but they also crack open nuts and bones for the nutritious marrow inside. The first stone tools were manufactured around 2.6 Mya with evidence from the Kada Gona site in the Hadar Region of Ethiopia. Other early sites include Omo and Bouri in Ethiopia and Olduvai Gorge, Tanzania.

What hominin species made Oldowan tools? It is certain that early *Homo* was making them but did they invent them or did australopiths? And did *Paranthropus* make them? The obvious benefit of stone tools would be to use them to butcher animal bones for meat and bash the bones open for the nutrient-rich marrow inside. We usually assume that only early *Homo* was eating meat. But they would also be useful to australopiths and *Paranthropus* for cracking open nuts and for digging up tubers and other underground foods. As of yet, there is still no clear-cut evidence to know one way or the other.

Building on the simple Oldowan technology, the Acheulean industry that arose around 1.6 Mya included more sophisticated tools including handaxes and picks. These tools were still crude but looked more symmetrical than previous ones. They are called "bifaces" because the stone is flaked-off and fashioned on both sides to give a symmetrical appearance. As opposed to much of the Oldowan Industry, even a nonexpert can clearly surmise that Acheulean tools took serious skill to produce.

The Acheulean tradition, named for the site of its first discovery in France, lasted for up to 1.5 million years in some localities. Clearly the tools in the Acheulean kit were useful for *H. erectus* and Archaic humans for processing foods. For many years it was thought that the advent of Acheulean technology allowed *H. erectus* to radiate successfully outside of Africa. However, other animals like elephants also originated in Africa and dispersed across the Old World without the aid of stone tools. Furthermore, Dmanisi, the oldest hominin site outside of Africa that has preserved tools, has only Oldowan tools. There is a general lack of Acheulean tools at Asian sites, a trend that was first seriously contemplated by archaeologist Hallum L. Movius. He drew a theoretical threshold line ("Movius line") across Eurasia beyond which no Acheulean tools were found in the East. There are many reasons Acheulean could be sparse in Asia and Indonesia. It is quite possible that hominins dispersed out of Africa before Acheulean technology was invented. Also, eastern hominins could have relied more heavily on organic tools made from things like bamboo, which did not preserve. Movius's line may not last much longer, however, because in 2000 archaeologists reported that they discovered handaxe-like tools (nearly indistinguishable from African ones) in the Bose Basin of China dating to 800 Kya.

By the time Archaic humans emerge in the Middle Paleolithic they show an improvement on the Acheulean that involves the use of the Levallois technique which is a method used to control flake size and shape. By preparing the core first more flakes, sharper flakes, and longer flakes can be produced from a single core. It is a process that requires several steps and is indicative of higher intelligence.

Neanderthals, for the most part, are credited with making Mousterian tools. They used Levallois and other prepared core techniques. The tools show increased variety, required soft hammers like bones and antlers to make, are often retouched or reflaked and reshaped to sharpen and hone them. Clearly brains and not brawn are needed to make Mousterian tools. Some Neanderthals moved on to making more advanced Châtelperronian tools that are very similar, if not identical, to some modern human tool kits from the same time period.

The emergence of any particular hominin species does not correlate with the concomitant emergence of a particular stone tool industry. For instance, *H. erectus* fossils appear before Acheulean tools do and the so-called revolutionary modern human/Upper Paleolithic tool kit does not appear until about 150,000 years after the first human fossils.

By 45 Kya, humans were making blades (very long flakes) and microliths (small flakes that were also hafted onto other tools). The earliest Upper Paleolithic tool culture, called the Aurignacian, superseded the Neanderthal cultures everywhere by 28 Kya. Like other aspects of human behavior, humans took toolmaking to the extreme. They invented the most tool types, used diverse materials (stone, bone, horn, antlers, and teeth), procured raw materials from up to hundreds of miles away, and also produced the smallest tools on record.

ANATOMICALLY MODERN HUMANS

Anatomically modern *H. sapiens* first arose in Africa and then eventually spread to nearly every corner of the earth, capable of surviving in all climates and environments. The second major dispersal of hominins, undertaken by modern humans, was a wider dispersal than that of the initial dispersal by *H. erectus*. Modern humans appear in Europe around 40 Kya, in Australia between 60 and 50 Kya, made it across the Bering Strait from Russia to North America some time around 30 Kya, and almost immediately appear in South America after that (Figure 6.1).

Several skeletons found at the Cro-Magnon rock shelter site at Les Eyzies in France, date to around 40 Kya. It is because of this notable discovery by Louis Lartet in 1868 that we adopted the nickname "Cro-Magnons" for fossils of our species.

Until recently the oldest modern human fossils came from the Herto site in the Afar Region of Ethiopia, dating to about 160 Kya. The three individuals, including two skulls, still retain small browridges and other primitive features so the discoverers gave them a subspecies *Homo sapiens idaltu* meaning "elder." One of the skulls had cut marks and was polished probably from being carried around after the individual died, perhaps indicating an early stage of ritual behavior.

Now that the site has been redated to 195 Kya, the very earliest fossils of fully modern humans on record are two skeletons, Omo I and II, which were found near Ethiopia's Omo River, right across from the northern border of Kenya. These fossils confirm molecular clock estimates which point to an East African origin for our species at around 200 Kya (see Chapter 4).

Although there is variation among and within modern populations, all humans share cranial features that distinguish us from earlier hominins. These traits include high round skulls, vertical flat foreheads with small or no browridges, chins, and an average brain size of 1,350 cc—slightly smaller on average per body size than Neanderthals, but larger than any other hominins. There is little to no postorbital constriction meaning the muscles for chewing are extremely reduced and the brain is expanded in their place. The small jaw has a prominent chin and small teeth. The muscle markings on the skull and the entire skeleton are less robust and the bones are more gracile or slender and weak than previous hominins. The face is tucked under the skull and is less prognathic than in previous hominins. The earliest humans also had tropical body proportions (see Chapter 5).

Once modern humans emerged they filled tropical, temperate, and arctic niches almost immediately. The first humans were foragers, gatherers, hunters of large and small prey, and fishermen. They traveled far distances to obtain raw materials or to trade for them. Their behavior and culture is characterized by complexity, flexibility, and innovation. For the first time, bones were being made into needles (oldest are 26 Kya), awls for sewing clothing, and spear-throwers (or "atlatls"), which add incredible velocity and distance to a thrown spear. Beads, pendants, and other evidence of personal adornment are everywhere by the time modern humans take over in the Upper Paleolithic or Late Stone Age.

HOMO FLORESIENSIS

As is evident by the Pleistocene ape *Gigantopithecus*, a creature like "Bigfoot" probably did exist in the past and fossils from Indonesia suggest that "hobbits" did too.

In 2003, a team of Australian and Indonesian researchers led by Peter Brown dug down into the floor of a cave called Liang Bua and uncovered a skull with an associated skeleton (LB 1) and the skeletal remains of at least eight other small individuals which were nicknamed "hobbits." Only some of the remains have so far been published but they are currently the most controversial fossils known to paleoanthropology.

These diminutive hominins stood about 3 feet tall and had brains the size of chimpanzees. The physical description may sound like an australopith, but complex stone tools were associated with them in the cave and there is evidence that they hunted pygmy elephants. Also, the site dates between 18 and 13 Kya, which is far later than any australopiths, and the morphology of the teeth and skull resemble early *Homo*. So the fossils were given a new species, *Homo floresiensis.*

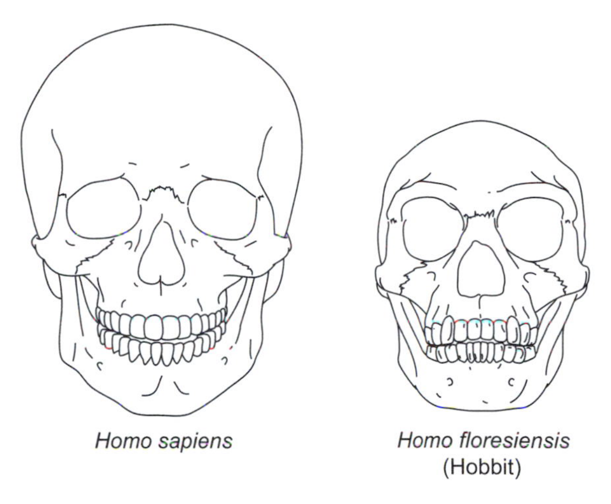

Figure 3.16 A comparison between the skulls of a modern human (*H. sapiens*) and a so-called "hobbit" (*H. floresiensis*) shows how they differ in their faces and overall sizes. *Illustration by Jeff Dixon.*

Reminiscent of the early assessments of Neanderthals, critics argue that the small skull (380 cc) is pathological (Figure 3.16). They posit that the hominin had a condition like microcephaly where the brain does not develop properly and remains dangerously small into adulthood. Supporters of the new species and its "normalcy" point out that microcephalics often have a misshapen cranium and face and the Liang Bua skull does not. In fact, the LB1 endocast shows that although it was small, the *H. floresiensis* brain was sophisticated. Two different genes (*microcephalin* and *ASPM*) when mutated, can cause microcephaly, but unfortunately, even if researchers were able to extract ancient DNA from *H. floresiensis*, the chances are very slim that a sequence containing either of these genes would be preserved.

Although there is only one skull so far, there are several postcranial elements that share the small body size. *H. floresiensis* looks like a dwarfed species of hominin, although it does not share limb proportions with living dwarfed or pygmy humans. Dwarfing is common for large-bodied animals like elephants and hippos that evolve in isolation on islands. At present, *H. floresiensis* is seen as a descendent of Indonesian *H. erectus* that got isolated on Flores 800 Kya and succumbed to the ecological pressures of island living by dwarfing its body size. Further research is needed to understand how the mammalian brain, particularly the hominin brain, reacts to island dwarfism in order to best address the microcephaly and the dwarfing hypotheses and also to determine if *H. floresiensis* is indeed a descendent of *H. erectus.*

4

Modern Evidence

MOLECULAR CLOCKS

DNA is not just the basic building block of life; it is also a data recorder. The genetic sequence of an individual tells the tale of their geographic ancestry and the genetic code of a species tells the tale of its evolution. By comparing the DNA sequences of different organisms, it is possible to determine how closely related those two organisms are, and it is also possible to estimate the time since they shared a common ancestor. Similarities in DNA decrease with the decrease in relatedness at the individual level, at the species level, and so on. Identical twins are born with identical genomes. Normal mammalian siblings share an average of 50 percent of their DNA, just like offspring share an average of 50 percent of their DNA with their parents. Grandchildren share 25 percent of their DNA with their grandparents. And so on. Whether two individuals shared a great-great-great-grandmother 100 years ago or whether they shared an ancestral root 100 million years ago, it is all spelled out in their DNA.

The method for comparing two sequences and then estimating the time since their common ancestry is based on the principle known as the "molecular clock." Late in the 1960s, Vincent Sarich and Allan Wilson were pioneers in applying the concept to human evolution. Their initial analysis comparing the African apes and humans resulted in a tree with humans and African apes grouped together, excluding Asian apes. This result along with corroboration from many other laboratories, contradicted the assumption based on anatomy and behavior held by many scientists for most of the first half of the 20th century. The long-standing view had grouped Asian and African apes together to the exclusion of humans, but these new genetic comparisons showed that humans were more closely related to gorillas and chimpanzees than African apes were to Asian apes.

The estimated splitting dates within the apes were also changed once molecular clocks were invented. Sarich and Wilson placed the chimpanzee–human split around 5 Mya and since then an abundance of molecular clock research has produced similar estimates between 8 and 4 Mya. Before molecular clocks, the split time was assumed to be much deeper in prehistory, between 30 and 20 Mya, which misled some prior paleoanthropologists into searching in much too old rocks for fossils of our bipedal ancestors.

Basically, the genetic distance between two samples is assessed by directly comparing the nucleotides in equivalent positions on each of the two sequences and counting up the number of times they differ. Molecular clock methods can be applied to genes, gene families, and, theoretically, to whole genomes. They are also commonly applied to the DNA sequences located in the mitochondria, the so-called "powerhouses" of the cell because they produce energy. Although DNA in a cell's nucleus can be used for molecular clock analysis, mitochondrial DNA (mtDNA) is often preferred because mutations in mtDNA accumulate randomly and often with little consequence to the fitness of the organism (unless they affect the proper functioning of the mitochondria, leading to muscle disorders). Neutral noncoding regions are preferred for use in molecular clocks since without selection acting on them, mutations in these regions accumulate at a steady, clock-like rate per generation.

Using molecular clocks, the splitting times of all the major groups of primates have been estimated and verified. The estimations generally span a range of time but they tend to average about the same. (Figure 2.2 shows the splitting dates for major primate lineages.) Variation in estimated splitting times can be influenced by the calibration method used, or as discussed above, whether or not neutral sequences are used.

The clock is calibrated using fossils and this helps determine the time of divergence at each branching event in the tree thereafter. A good calibration point for the molecular clock analysis of ape and human splitting times is the fossil evidence for the divergence of Old World monkeys and hominoids at the Oligocene–Miocene boundary, 23 Mya.

DNA–DNA Hybridization

One way to compare genetic sequences is as follows. The DNA is heated up to break the bonds between the base pairs that hold the two strands together. The double-stranded DNA molecule splits into two separate strands in a process called "denaturing." Then the denatured DNA from the species to be compared is added to the mix and the entire sample is allowed to cool. During cooling, the DNA strands reassociate ("anneal" or "hybridize") to form new complete, double-stranded DNA molecules. The closer the species are, phylogenetically, the more similar their genetic sequences (base pairs) will be, the more likely their DNA strands will join together during this process, and the tighter their bonds will be. In order to evaluate how well the sequences fit together, they are reheated. If they denature faster than the original DNA molecule (from a single species), then there is a weaker bond because less base pairs are matched. If they require an equivalent amount of time to split as the original DNA, then the sequences are basically the same.

HUMANS AND CHIMPANZEES: THE NARROW DIVIDE

Ever since Emile Zuckerkandl and Morris Goodman independently compared blood proteins of African apes and humans and determined they were indistinguishable, the remarkable molecular similarities between chimpanzees and humans have been obvious. But now with the sequencing and mapping (pinpointing the location of genes on chromosomes) of whole genomes of chimpanzees and humans, it is possible to identify the genes responsible for the expression of the similarities like those blood proteins and also the differences like hair patterns and even brain size.

Aside from humans and chimpanzees, many genomes have been sequenced and include, for example, the mouse, *E. coli*, puffer fish, honeybee, silk moth, and rice. Those of gorillas, orangutans, bonobos, and rhesus monkeys are under construction. So far, as one would predict, animals that look similar have similar genomes. Perhaps the most humbling find from the mapping of the human genome is that humans have between 20 and 25,000 genes, which is only about twice that of a worm or fruit fly.

Of course, for investigating human origins we are most interested in how humans and chimpanzees matchup. Sequence comparisons have consistently shown that humans and chimpanzees are the most closely related hominoids sharing about 99 percent of their genetic code. The genomes prove unequivocally that chimpanzees are our closest living relatives, just as Darwin predicted and just as numerous other studies indicated that led up to the whole genome analyses.

Depending upon how one looks at the genomes of humans and chimpanzees, the difference in DNA can be estimated to be larger than 1 percent difference. Some estimates show up to an 8 percent difference. Straight nucleotide base pair differences are very small, on the order of less than 2 percent difference, but if the lens is pulled back and entire genes (segments of base pairs) or gene families (genes that get inherited together or that work together) are considered, then the human–chimpanzee genetic differences can be larger. This makes sense since most of the raw material is the same between chimpanzees and humans, it is just tweaked differently in each one. Both have hair, but it is just differently grown and arranged. Both have arms and legs, but they are just different lengths. Both have larger than normal brains and greater than average intelligence but humans are just more exaggerated. With hardly any sequence differences from chimpanzees, humans are not so derived as was previously assumed and we fit even more comfortably in the Tree of Life.

But humans are clearly different from chimpanzees, so what makes a human not a chimpanzee and conversely, what makes a chimpanzee not a human? Can we find the genes for big brains, language, tooth size, etc? Scientists are comparing, and will be comparing forevermore, the chimpanzee and human genomes to find regions that differ. During a scan of the sequence on a computer, for instance, a scientist may find a particularly interesting region that separates chimpanzees from humans. They will then look to databases of mapped genes in other animals to see if those genes have been identified and if their functions are known. If not, they may knock a gene down (a "candidate gene") in a mouse to see what effects, if any, are visible in the phenotype from the loss of the gene's function. Such an experiment in a model animal helps determine the function of genes since mice have very similar genomes to chimpanzees and humans.

Out of all the chromosomes, the Y chromosome seems to bear many of the differences between chimpanzees and humans. About 30 percent of our proteins are the same and those that differ are only separated by a couple of amino acid changes. Many of the types of genes that have experienced more evolution in humans than in chimpanzees—which is tracked by number sequence changes compared to primitive genomes like mice—are involved in immunity, sperm and egg production, and sensory perception like smelling.

Comparisons between the chimpanzee and human genomes mean nothing without an out-group to help decipher what traits are primitive (shared with the out-group) and what traits are derived (unique to the species). Once the bonobo, gorilla, orangutan, and rhesus monkey genomes are complete it will be possible to tell which genes arose on the chimpanzee lineage and which arose on the human lineage. Having the genomes of closely related out-groups mapped will lead scientists toward the genes that distinguish humans from the other apes.

The practical and promising aspects of whole genome sequencing and mapping involve finding cures for diseases and genetic disorders, but sequencing of whole genomes is vastly improving our understanding of the relationships between genotypes and phenotypes and the complicated biology of inheritance. Looking across genomes, especially at regulatory genes, has shown us that Mendelian inheritance—a biology driven by single genes and discrete traits—is the exception rather than the rule. In reality, many traits are polygenic where different alleles on more than one gene, often on different chromosomes all together, are responsible for a single trait. Also, traits are more likely to be continuous than discrete, ranging across shades of colors, for example, as opposed to being simply black or white. Additive effects are common, in which case the sum of the products of each involved gene results in the phenotype.

It appears that most of the differences between the DNA of humans and chimpanzees have to do with the activity levels of genes rather than the genes themselves. So, regulatory genes that either regulate expression themselves, or express proteins that regulate the activity of other genes, are where the human–chimpanzee variation occurs. A good explanation for this phenomenon could be that turning a gene on or off, or changing the levels of proteins expressed, is evolutionarily easier than changing the genes, amino acids, or proteins themselves.

"JUNK" DNA

The function of over 95 percent of our DNA is still a mystery. That is, we have spelled out the code, but have discovered that most of it does not code for proteins. Genes can be separated by a vast desert of noncoding DNA, which is sometimes called "junk" DNA. But is it useless? Probably not, because included among noncoding sequences are the crucial promoter regions which control when genes are turned on or off.

The human genome has more noncoding DNA than any other animal known to date and it is not clear why. At least half of the noncoding sequence is made up of recognizable repeated sequences, some of which were inserted by viruses in the past. These repeats may provide some genomic wiggle room. That is, long stretches of noncoding DNA provide a playground for evolution. It may be a huge selective advantage to have all that raw material available to mutate and either modify existing traits and behaviors or express new ones all together. Humans are characterized by the ability to be flexible and to adapt quickly, so our junk DNA is potentially a priceless contribution to our humanness.

An intriguing difference in chimpanzees and humans occurs with the *MYH16* gene. A mutation, estimated to have occurred about 2.4 Mya, rendered it inactive in humans but it continues to function well in chimpanzees, rhesus monkeys (*Macaca mulatta*), and other primates. *MYH16* is involved in the development of jaw muscles for chewing so scientists hypothesize that the loss of function of this gene is linked to the drastic change in skull morphology in early members of the genus *Homo* around 2 Mya. Hominins during that transition time clearly had smaller teeth and chewing muscles and were in the midst of adopting a carnivorous way of life where stone tools began to take much of the functional burden away from the teeth and jaws. Although *MYH16* is probably not the only gene involved in the chewing muscle complex (as it did not disappear completely with the mutation), it could definitely reveal a significant event in our evolution.

The mapping of whole genomes has revealed how difficult it is to find the "gene for" most traits, not only because phenotypes are expressed through complicated orchestrations of genes, but also because they can be heavily influenced by the environment during development and during life. When looking for specific genes, geneticists can control for environmental factors by observing identical twins that have essentially identical genotypes but grew up in different environments (i.e., separate bodies). Unfortunately such experiments are impossible to perform with *Homo erectus*, but understanding the environmental influences on trait development can help paleontologists decipher which traits are important for identifying species and tracking evolution and which traits are useless for those purposes because they are easily changed by nutrition or other nongenetic, environmental factors. An important area of genetic study uses quantitative trait loci or QTLs to determine which skeletal and dental traits of fossils are most informative for understanding evolutionary history, as compared to those that are too easily changed by the environment during life.

The few differences between humans and the great apes are actually the genetic foundation for our rapid cultural evolution and geographic expansion after 200 Kya. Only a few genetic changes were necessary to make human prehistory and history possible.

MITOCHONDRIAL EVE AND Y-CHROMOSOME ADAM

Geneticists have traced similarities in our mitochondrial DNA back to a single shared mother of us all and have also determined the hypothetical father of all the males on Earth today by tracing back through their Y-chromosome variation.

Most people have only heard of mitochondrial DNA because of the sensation "mitochondrial Eve" caused in the 1990s. mtDNA is not the same as nuclear DNA, which is found in the nucleus of the cell. MtDNA is located outside of the nucleus in energy-producing organelles called mitochondria. For several reasons mtDNA is an excellent source of evolutionary information. First of all the genome for mtDNA is smaller than nuclear DNA and it is therefore easier to analyze than nuclear DNA. Second, each cell has thousands more copies of mtDNA than copies of nuclear DNA so it is more amenable to work with in the laboratory. Third, mtDNA does not accumulate mutations as fast as nuclear DNA. And finally, mtDNA is only inherited maternally, meaning that everyone receives their mtDNA from their mother. There is no recombination with the father's mtDNA because the mtDNA in sperm are destroyed at the time of egg fertilization. Because of the hereditary properties of mtDNA, we can trace every living person's genes along a maternal lineage back to a single mother of all modern human mtDNA. This ancestor is known as "mitochondrial Eve."

Rebecca Cann and her colleagues were the first to estimate a common mtDNA ancestor for all living humans. Since then others have tried to answer the same question and have reported varying but overall similar results. Based on comparing sequences between different populations across the globe, they concluded that we share a common mtDNA ancestor who lived about 200 Kya in sub-Saharan Africa.

Of populations all over the world, the African samples show the most variation in the mtDNA genome. In other words, the populations outside of Africa only exhibit a subset of the variation seen in the African sample. According to molecular clock theory, the sequence with the most variation caused by mutations has had the longest time to evolve or to accumulate those mutations. MtDNA analysis confirms that people have been evolving in Africa for a longer portion of human evolutionary history than anywhere else.

Bottleneck

MtDNA and Y-chromosome studies as well as those on nuclear genes support a population expansion out of Africa beginning 100 Kya. Through statistics and computer simulations based on sequence analyses of different populations around the world, geneticists from many different laboratories have concluded that *Homo sapiens* experienced a bottleneck between 150 and 50 Kya. At that point, human variation was particularly low. Studies of genetic variation within and between populations suggest that at some point, the global human population could have been as small as 10,000, which would have been an endangered level, on par with bonobos and mountain gorillas today. Since that loss of genotypic diversity at the bottleneck, humans have experienced phenotypic diversification and have developed regional variation to make us the polytypic or variable species we are today.

Of course, "Eve" was not the only living hominin female at the time. Other women were alive and reproducing. The special point of interest about our Eve is that she is the *only* woman whose entire descending lineage had surviving and reproducing females in every generation. All the other lineages that were plodding along 200,000 years ago have since ended and contributed nothing to our mtDNA.

By comparing worldwide variation at the Y chromosome, geneticists were able to corroborate the mitochondrial Eve hypothesis. Since there is no female counterpart to the Y chromosome it, like mtDNA, does not undergo recombination. So the Y chromosome is passed, as is, from father to son. All the world's human Y chromosomes converge on a "Y-Chromosome Adam" who lived about 100 Kya. And just like mtDNA, because African Y chromosomes show the most variation, it follows that those lineages have been evolving the longest and that the human Y chromosome originated there.

The recent age of our common ancestors at 200 Kya and 100 Kya is strong evidence against the Multiregional model for human origins and supports the Out of Africa model, which is also called the Garden of Eden hypothesis (see Chapter 6). Mitochondrial Eve and Y-chromosome Adam should not be taken literally as if they are the biblical couple. According to strict interpretation of the Bible, Adam and Eve were the very first humans on earth. However, mitochondrial Eve and Y-chromosome Adam are consistent with evolutionary theory which assumes that they had ancestors and contemporaries. They are simply the only woman and the only man to have descendents that survived to the present. It is important to remember that Y-chromosome Adam and mitochondrial Eve need not have lived at the same time. They each signify the roots of different aspects of our biology, which evolution has been building upon for eons.

The gene pool of mtDNA and Y chromosomes in Africa contains more variation than anywhere else and the genetic variation outside of Africa represents only a subset of that found within the African continent. Therefore, from a genetic perspective all humans are Africans. Because of their tropical habitat in which they would have experienced high levels of UV radiation, "Adam" and "Eve" probably had darkly pigmented skin.

ULCERS

The genetic root of all the strains of human gut bacteria (*Helicobacter pylori*) is located in Africa. About every other human carries *H. pylori*, that is, the bacteria responsible for stomach pains and ulcers and is the only known microorganism that can survive in such a highly acidic environment. Based on genetic differences among populations' strains of the pathogen, the initial *H. pylori* infection in humans is estimated to have occurred around 60 Kya. Because the sequences of the African strains of the bacteria are more diverse, it is assumed that evolution has had longer to act in Africa, so the ancestral root is located there. The date of 60 Kya implies a later "Out of Africa" scenario than that constructed by the evidence for mitochondrial Eve since humans carried *H. pylori* out of Africa with them.

Molecular phylogenetics also show that *H. pylori* is a distant relative of a similar ulcer-causing species of bacteria in lions, tigers, and cheetahs. The genomes of these microbes show that some of the genes in the human form are still functioning while their counterparts in the feline form (*Helicobacter acinonychis*) are not. Since it is unlikely (i.e., not parsimonious) that these genes would have been lost and then restored in the human form, it is probable that the human form, rather than the big cat form, is more like the ancestral microbe. Therefore, it is quite possible that the bacteria was transmitted from humans to big cats and, based on molecular clock estimates of the two species' sequences, the transmission occurred 200 Kya. Big cats could have contracted the microbe after eating a human and then spread it to other cats the way humans spread it to one another through close physical contact.

HUMAN ADAPTATION

Although the genes involved in the expression of the complex skeletal and dental traits we track through the hominin fossil record are still being identified, examples of selection-driven adaptation in living humans illuminate the recent and continuing evolution within our species, and exemplify the speed with which evolution can transform human populations.

The maintenance of a harmful allele like that for sickle cell trait shows how humans can adapt with their own biology to fight diseases. The recent evolution of lactose tolerance into adulthood in different populations shows how quickly dietary adaptations can evolve and also how readily convergent evolution occurs. Selection of varying amounts of skin pigmentation in humans is a striking example of how humans adapt to particular environments.

Sickle Cell Trait

A gene with two or more alleles, each at an appreciable frequency, is a genetic polymorphism. In some African populations, the gene for the morphology of red blood cells is one such case. One allele (which we will call "A") codes for expression of normal red blood cells, but the sickle allele (which we will call "S") codes for sickle-shaped red blood cells. The S allele is caused by a point mutation changing only one amino acid. The two alleles are codominant, so neither masks the expression of the other. Individuals with two sickle alleles who are homozygous (SS) are affected with sickle cell anemia, which is a lethal disease where sickle-shaped red blood cells are essentially rendered handicapped and cannot perform their oxygen transport duties properly. Because of the negative effects of the S allele, one would expect it to exist in small frequencies, but it can reach frequencies of 20 percent or higher in populations in tropical regions of Africa, Southeast Asia, and Indonesia where malaria is prevalent.

Populations living in malarial infested regions have evolved a natural means of resistance to malaria using the S allele. Noncarriers (AA) have normal red blood cells but die more often from malaria than sickle allele carriers (AS) who are resistant to malaria. The malaria parasite (various species of the *Plasmodium*), which lives inside red blood cells, actually induces sickle-shaped red blood cells to rupture, so the parasite cannot eat, survive, and reproduce. The sickle allele carriers have some sickle-shaped cells but not enough to cause severe health problems or death. However, people with the homozygous genotype (SS) will most likely die from sickle cell anemia. Africans carrying the sickle allele who were taken to America in the slave trade no longer had an adaptive use for it in a land without malaria, so the successful adaptation in Africa is now a maladaptation in America. The sickle cell story is a powerful lesson that genes can be beneficial or harmful depending on the environment.

Lactose Tolerance

The ability to digest lactose into adulthood is a relatively new adaptation in humans that arose concomitantly with the advent of agriculture about 10,000 years ago.

Lactose is a component of milk and dairy products. Lactase is the enzyme that breaks down lactose during digestion so it can be absorbed by the gut. Normal or "wild type" humans have the ancestral condition, which is to stop producing lactase at about twelve years of age. They are lactose malabsorbers, or lactose intolerant, like other adult mammals.

However, some European and African groups that have long histories of pastoralism and animal domestication continue to produce lactase into adulthood and retain the ability to digest lactose. Early humans were strictly hunters and gatherers until about 10 Kya when some populations adapted new ways to obtain food: using domestication to keep the food source close. Selection pressures were strong on those people who were able to maximize these animal resources through digesting their dairy products throughout their lives. The ability to digest lactose as an adult must have been a huge nutritional advantage in these populations because selection favored it strongly. For example, nearly all Dutch and Swedish adults are lactose tolerant.

Genetic analyses by Sarah Tishkoff and others have shown that African and European groups converged on the same adaptation independently. Each group uses a different variant of a regulatory gene (i.e., a slightly different variant of an allele called an "SNP" for single nucleotide polymorphism) to control the gene for lactase production, which is called *LCT*. So far there are four different mutations that keep the lactase gene switched on. Each allele occurs in much higher frequency in populations that have long histories of dependency on domestic dairy animals: (1) Dutch and Swedes who are related to the ancient "Funnel Beaker" cattle-raising people of north-central Europe, (2) Nilo-Saharan-speaking groups in Kenya and Tanzania, (3) the Beja people of northeastern Sudan, and (4) Afro-Asiatic-speaking groups living in northern Kenya. The evolutionary tale of lactose tolerance is a powerful one because it highlights the influence of culture on biological evolution. Even more importantly, lactose digestion is a shining example of how quickly the biology of humans can adapt in order to survive better.

Skin Color

Because of the wide spectrum of variation, skin color is probably the most common and most conspicuous trait humans use to identify major phenotypic differences in others. However, Nina Jablonski and her colleagues have shown that there is a strong adaptive function behind human skin color variation.

Our closest relatives do not have skin color variation like ours. Chimpanzees have pale skin underneath their covering of black fur. Young chimpanzees have light faces, but with age and years of sun exposure adult chimpanzee faces develop dark spots or become dark all over. Based on parsimony, we assume our last common ancestor with chimpanzees was also pale-skinned with dark fur.

Regardless of how or when the drastic body hair reduction in humans occurred (probably with the emergence of the genus *Homo* at about 2 Mya), it is highly probable that darkly pigmented skin underwent very fast, very strong positive selection once hominins lost their furry coat. Pale naked skin is a great health risk under the equatorial African sun, so darkly pigmented skin would have been strongly favored.

Why is darkly pigmented skin better for people in the tropics? Melanin, the substance that gives color to skin, regulates the penetration of harmful ultraviolet radiation (UVR) from the sun. The more melanin in the skin, the more it can act as a natural sunblock as there is more solar radiation in the tropics than anywhere else in the world. The risks of sunburn and skin cancer are greatly reduced in people with dark pigmentation. One type of UVR, UVA, damages cells and DNA. It causes the breakdown of folate (a form of folic acid), retarding DNA replication and cell division in embryos that can then lead to their spontaneous natural abortion. This is a direct fitness affect of UVR on humans, since it negatively affects successful reproduction.

If dark skin is advantageous, why do some people have light skin? In regions further from the sunny tropics, darkly pigmented skin is actually maladaptive because it is too effective at blocking UVB (another type of UVR). UVB penetration of the skin is necessary for humans to metabolize vitamin D in their skin. At high latitudes away from the equator there is much less UVB present. Having less pigment in the skin allows more UVB to penetrate and permits the necessary metabolism of vitamin D (which is necessary for the body to absorb calcium, among other functions). In regions with significantly less UVB in the atmosphere, humans lost much of their skin pigmentation under the selective pressure to facilitate this process.

The consequences of vitamin D depletion include the onset of rickets, where bones do not form properly, often leaving the affected individual incapable of independent locomotion. A person's leg bones with rickets literally buckle under their weight. What's more, women who suffer severely from rickets have misshapen pelvic bones that render natural childbirth difficult to impossible. This is a direct link to fitness. In fact there are very few areas away from the tropics where people with darkly pigmented skin can live year-round and absorb enough UVB. Many people of African descent living in the United Kingdom, for example, must supplement their diets with vitamin D rich food like fish in order to live healthily there.

Molecular clock analysis of the *MC1R* gene associated with skin pigmentation points to a date of 1.7 Mya for the emergence of dark skin. It is quite possible that the people of Indonesia, Australia, and the South Pacific evolved dark skin separately, from an ancestral Asian stock that had reduced much of their pigmentation. And it is also possible that light skin evolved separately in Europe and in Asia. Geneticists will soon have the answers to these questions.

ANCIENT DNA AND THE NEANDERTHAL GENOME

When the preservation conditions are right, it is possible to extract DNA from ancient—even fossilized—animal and plant remains. Cold dry environments or wet anaerobic ones are usually the best for preserving ancient DNA, or aDNA.

Because the survival of old molecules of DNA is rare, and because those that survive actually degrade over time, there is a general rule that specimens older than 100,000 years old are not viable candidates for aDNA extraction. (A potential exception may be 400,000-year-old plants frozen in Siberian ice that may have preserved aDNA.)

Degraded aDNA sequences can be compared to closely related modern species to correct for sequencing errors that occur when nucleotides degrade to "look" like different nucleotides. Cloning an organism from aDNA to bring it to life could only be possible if the entire DNA is preserved and the chances of this occurring are very slim.

So far aDNA has been extracted from at least six Neanderthals (from Germany, Croatia, Georgia, Belgium, France, and Spain) and at least five modern human fossils (from Czech Republic, France, and Italy). MtDNA is useful for aDNA studies because there are thousands of mitochondria containing mtDNA in a cell as opposed to just one nucleus with DNA per cell. After degradation, mtDNA has a much better chance at preserving than nuclear DNA. Small portions of the bones or teeth must be drilled out and ground up in order to be analyzed, but fortunately casting technology is at the point where high-quality replicas of the specimens can be crafted prior to aDNA extraction. It is a destructive process so paleoanthropologists are constantly weighing the merits of preserving the fossils and their morphology on one hand versus extracting the precious genetic information they possess.

Ancient DNA Methods

Before aDNA can be sequenced it must be cloned and amplified using a technique called PCR (polymerase chain reaction), which is the same technique used in paternity testing. Thousands of copies of the aDNA are needed to perform the laboratory techniques used in sequencing and PCR provides those copies. All aDNA analyses must take place in a physically isolated work area to avoid human contamination, which can result in the amplification of DNA that is not the aDNA being investigated. In order to detect sparse aDNA, multiple extractions from the fossil and multiple PCR procedures must be performed. An act as simple as handling a fossil without gloves can contaminate a specimen, so aDNA analyses of fossils that have been handled for many years in museum collections are treated with extra care. Because of all the chances for human contamination and because it may be very difficult to determine whether or not the aDNA or the geneticist's DNA is being amplified and sequenced, extra precautions are necessary. For instance, results should be repeatable from the same, and different aDNA extractions of a fossil specimen and separate samples of a specimen should be extracted and sequenced in independent laboratories.

Based on ancient mtDNA analyses we know that Neanderthals are three to four times more different from modern humans than modern humans are from one another. Plus, the genetic variation between Neanderthals and modern humans is much greater than that within modern humans alone. The variation among Neanderthal aDNA from fossils at distant sites is similar to that among modern human populations.

Neanderthals fall between chimps and humans and outside living and fossil modern human range. Based on molecular clock rates of mitochondrial aDNA, the Neanderthal lineage diverged about 500 Kya from the lineage that led to modern humans. Neanderthals, according to their genes, are distinct from modern humans and they are not more closely related to modern Europeans than any other modern human (which is a strike against the Multiregional hypothesis for human origins, see Chapter 6). It is therefore unlikely we are descended from Neanderthals and unlikely we shared genes with them (i.e., interbred with them). We should probably consider Neanderthals our distant cousins.

Methods of aDNA analysis have advanced far enough to allow scientists, like Svante Pääbo, to extract nuclear DNA from Neanderthals. With enough aDNA from enough specimens, eventually large portions of the Neanderthal genome will be reconstructed. So far, the 38 Kya male found in Vindija Cave, Croatia, produced a sequence of around a million base pairs of nuclear DNA, which is around 0.03 percent of the genome.

With nuclear DNA, the functional genes and the sex chromosomes can be analyzed. So for instance, the Y chromosome of Neanderthals is vastly different from those of humans and chimpanzees compared to other chromosomes, suggesting that little interbreeding occurred at least by the latest Neanderthal times.

Neanderthals share an estimated 99.5 percent of their genome with humans, but humans are only 0.1 percent different from each other. So since variation does not overlap, the separate species designation of Neanderthals and humans is supported. Like the mtDNA studies, the nuclear genome of Neanderthals shares no derived alleles that are special to European populations, meaning that Neanderthals did not contribute to European human evolution.

It will soon be possible to study functional and diseased genes in the extinct species. It will also be possible to test whether Neanderthals contributed to the human genome as some scientists have hypothesized is the case with the *microcephalin* gene (see Chapter 5).

The most promising aspect of the Neanderthal genome project is the opportunity to look for special genes in the human genome for language and art—traits and behaviors that are unique to humans and are not always considered to have existed in Neanderthals. Candidates for these genes would be those shared by chimpanzees and Neanderthals to the exclusion of humans. What's more, genes that are shared by Neanderthals and modern humans, but not chimpanzees, may point to genetic changes that occurred earlier in hominin evolution.

5

Interpreting the Evidence

BIG BRAINS AND INTELLIGENCE

The large human brain evolved relatively late in hominin evolution, once *Homo erectus* arrived on the scene. However, because the human brain is seen as the champion of human evolution, we will consider it first. The irrepressible curiosity surrounding human brain evolution is perpetuated by the very matter that is so puzzling. However, there is nothing inside the human skull that is unique (Figure 5.1). Only the relative sizes of the anatomical regions within the brain and the number of neurons and the nature of their networks are unique to humans. Primates in general have large brains compared to most mammals. In fact, Neanderthals even had bigger brains than we do, so it is the wiring of our brains (the number and nature of the neural networks), not the size of our brains, which sets humans apart.

There is a specific type of "spindle" neuron in the human brain that is only known to exist in the brains of great apes and cetaceans (an order of mammals that includes dolphins and whales). Like most of the ocean floor, there is much uncharted territory in the brain. It is still not exactly known what spindle neurons do and how (or if) they correlate to higher cognitive abilities. However, it is probably no coincidence that spindle neurons are shared by the groups that share complex social patterns, intricate communications skills, coalition formation, cooperation, cultural transmission, and tool use.

Brain size is correlated with intelligence at the species level. Humans have the largest brains for their body size, with cetaceans and great apes falling close behind, and these species are considered the most intelligent animals. Higher levels of mammalian intelligence are characterized by flexible problem solving in complex scenarios, incorporating

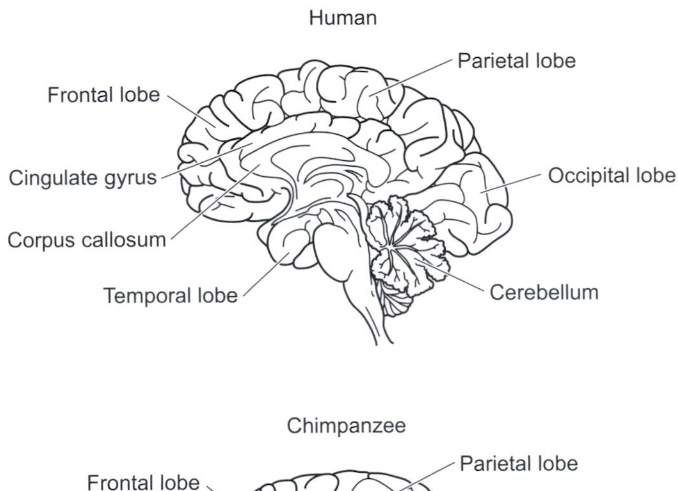

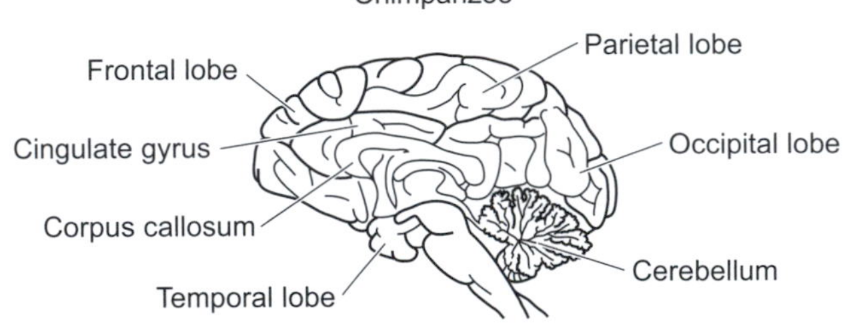

Figure 5.1 Cross-sectional slices through the brains of a human and a chimpanzee show their similarities in content but their differences in relative sizes of regions. *Illustration by Jeff Dixon, based on CT scans provided by Kristina Aldridge.*

novel solutions into existing behaviors, and curiosity. Human intelligence is distinguished by, for instance, our propensity for abstract and symbolic thought. The enlarged frontal lobe of humans in relation to the great apes is credited with advanced emotions, awareness, and memory. Strokes to parts of the frontal lobe can affect moral decision-making processes. The prefrontal region in particular is slightly increased in great apes and humans and contains centers for forethought and planning.

Much of what is known of ape intelligence comes from the laboratory where the animals are introduced to situations they do not experience in the wild. Chimpanzees will spontaneously draw and paint in captivity. At the Kyoto University in Japan, a chimpanzee named "Ai" learned over 100 visual symbols, including lexigrams, Japanese Kanji characters, letters of the alphabet, Arabic numerals, and also learned to understand some human speech and gestures. Many other chimpanzees and gorillas in captivity have learned to understand a huge vocabulary of English words in the context of complex sentence structures. In the wild, apes make and use tools. Chimpanzees are even known to save a tool to use later which requires forethought.

Within species, brain sizes vary and the correlation between brain size and intelligence does not hold. A person with a brain size on the lower end of the spectrum of variation is not necessarily any less intelligent than a person with the largest brain in the world. Measuring intelligence is also problematic. For example, intelligence quotient (IQ) tests applied to humans can only measure specific aspects of higher brain function, not the whole phenomenon. Physical intelligence, like the massive and complex neural coordination it takes to throw a baseball at 96 miles per hour into a catcher's mitt, is completely ignored by standard tests of intelligence.

Without being able to easily measure intelligence levels to compare *between* species, the measurable size of the brain is used as a proxy for amount of intelligence. And instead of using basic brain size, the size of the neocortex is compared, which is the region for spatial reasoning and sensory perception, and it is made up of the outer surface of the brain. Specifically, the ratio of the neocortex size to the size of the rest of the brain correlates to problem solving and flexibility or "intelligence" in primates. Robin Dunbar and colleagues have shown that neocortex size is related to social group size in primates (presumably because complex socialization requires exceptional intelligence).

The fossil record shows that the hominin brain clearly increases in size with time, but so does body size. Since brain size and body size are correlated across mammal species, values of brain size alone indicate nothing without taking into account body size. The encephalization quotient (EQ) of a species is the ratio of its brain size to its expected brain size based on the known correlation between brain size and body size in mammals. EQ offers a way of comparing brain sizes between vastly different-sized species (like whales and humans) and for tracking brain size increase in hominins (Table 5.1). A significant increase in brain size relative to body size occurred in *H. erectus* and continued thereafter until the emergence of modern humans.

Several different factors could have contributed to the selection for a larger brain in *H. erectus* and beyond. But first we should consider the evolution of primate intelligence. Major hypotheses for the evolution of primate intelligence are based on food procurement, food extraction, and socializing. If a food source is seasonal like fruits, a primate must

Table 5.1 Approximate Average Brain Sizes and Encephalization Quotients (EQ) of Living Hominoids and Fossil Hominins

Species	Average CC	EQ
Orangutan	400	1.1
Gorilla	505	1.1
Chimpanzee	385	1.5
Australopithecus afarensis	450	1.9
Australopithecus africanus	445	2.2
Paranthropus boisei	500	2.5
Homo habilis	630	2.7
Homo ergaster/ erectus	1,000	3.3
Archaic humans	1,300	3.5
Homo neanderthalensis	1,400	4.0
Human	1,450	4.3

mentally map sites, plan routes, and anticipate availability and quality of the resources. Such cognitive needs may also require traits like acute, stereoscopic, and color vision. Grey-cheeked mangabeys (a species of African monkey) have evolved the ability to use the weather to predict when figs will ripen. If the weather has been warm, they will return to a tree where they know figs are on the cusp of ripening. Extracting foods like hard nuts, roots, insect larvae, seed pods, and stems require higher cognitive functions as well as precise motor skills. Living in a group, like most primate species, involves competition, conflict, affiliation, cooperation, kin selection, and reciprocity, and tracking the behavior of others. Primate intelligence can and did evolve because of any of these models or combinations thereof.

Certainly these issues were at play in early hominins and all of them could explain the ratcheting up of brain size, but because human brains are larger and capable of higher cognitive functions than other primates, something else must have been at play. Brain tissue is very expensive to grow and to maintain. Developing such large brains would have required strong selection. An enhanced nutrient-rich diet, particularly rich in essential fatty acids provided by animal protein, would certainly help. It is probable that the incorporation of meat into the diet—which happened to coincide with body size and brain size increase during *H. erectus* times—contributed to brain size evolution. It is not concretely known why selection favored brain size increase, although it is easy to think of reasons to do with culture, diet, socialization, reproduction, etc. The genes *microcephalin* and *ASPM* likely played roles. *Microcephalin* is a critical regulator of brain development and mutations in the gene cause a disorder that is localized to the brain—microcephaly, which is severe brain size reduction on the order of up to 70 percent.

Functions of the brain are often localized to either the left or right hemisphere, which is called lateralization. Language is one such lateralized function with centers located in the left hemisphere. In several books beginning in 1983 and continuing to the present, William Calvin has proposed that the overwhelming trend for humans to be right-handed (controlled by left hemisphere) and also to be highly dexterous, is correlated to the buildup of neurons in the left hemisphere during language development in a feedback loop where language and handedness were each increasingly specialized with selection for complex neural networks. Calvin's hypothesis illuminates the significance of the evolution of physical intelligence (e.g., dexterity and hand use), which is often overshadowed by the evolution of mental intelligence (e.g., language).

EVOLUTIONARY PSYCHOLOGY

Have you ever read a diet advice column that suggests your inability to resist deep-fried fast food was actually adaptive in your evolutionary past? This is an example of the popular application of evolutionary psychology (EP), which is the blend of evolutionary biology and cognitive psychology.

Two of the most prominent evolutionary psychologists, John Tooby and Leda Cosmides, view the human mind as a Swiss army knife with hundreds or even thousands of "modules" that perform specific tasks. These modules, just like other parts of the body, evolved at different stages of hominin evolution between the LCA and prior to the advent of agriculture and civilizations around 10 Kya when all humans lived in small-scale foraging societies like contemporary hunter-gatherers. Examples of mind modules include those for food preferences (like fats and sugars), mating, theory of mind, predator avoidance, alliance formation, and language—all of which evolved to solve adaptive problems that were crucial for passing on genes.

For example, the theory of mind module allows humans to understand the mental states of others. Before reaching four and a half years of age, children cannot comprehend that others can hold different beliefs than them and as a result, they cannot lie convincingly. This uniquely human, but not learned, trait plays a crucial role in empathy, deception (of self and others), and manipulation in political/social situations.

Principles of EP are often applied to our eating and drinking habits and disorders. According to EP, humans have evolved food preference modules. These modules evolved for survival on the unforgiving African savannah and they govern the desire for certain foods and the dislike of harmful ones. Fat obtained from scavenging or hunting animals and sugar obtained from ripe fruits offer nutrients that would have been relatively scarce on the savannah. Both fat and sugar are needed for growth, maintenance, and metabolism (energy production).

Carcasses and fruiting trees are stages for between-species competition, and certainly would have been dangerous interaction zones between carnivores and hominins or between other primate species and hominins. In order to sustain a behavior, the benefits must outweigh the costs. So according to EP, fat and sugar preference modules evolved to make individuals desire the foods strongly enough to incite them to take risks.

There has been no observed selection for modern humans to lose the fat and sugar desiring modules even though many humans live in an environment where fat and sugar are readily available. Our evolutionary history has encountered a modern environmental mismatch. Now we run the risk of overeating fat and sugar to the point where obesity and heart disease are the number one killers of humans in industrialized countries. Binge-eating behaviors, which are considered eating disorders that lead to obesity and heart disease, could have been adaptive in the past if humans, like other large carnivores, were eating large meals spaced days apart.

One of the arguments over EP pits those that see everything as adaptations against those that see many traits as mere by-products of a limited number of adaptations. For example, the universal, cross-cultural belief in supernatural explanations for events or for a "higher power" is considered by some researchers to be due to a module that is hard-wired into all of our brains, but others consider it a spin-off from other functions involved in higher cognition and awareness.

BIPEDALISM

Humans are the only mammals to habitually walk and run on two legs. One of the million-dollar questions in human origins is why did a quadrupedal ape living at about 6 Mya start to walk bipedally? Four legs are far superior over two for strength, speed, and balance. Also, what kind of selection pressure was needed to evolve adaptations for bipedalism from quadrupedalism? The other question that goes hand in hand with the question of "why" is the question of "when"—when did hominins become habitually bipedal and lose most of their arboreal adaptations?

Several hypotheses have been offered to explain the origin of bipedalism, which is also usually considered the wedge that split the LCA into two groups: hominins and the lineage that led to chimpanzees. Many of them explain the maintenance of selection on bipedalism once it evolved but few explain its very origin or the impetus for its evolution.

Energetics may have played a role. Human bipedalism is efficient; it takes less energy for a human to walk upright than it does for a chimpanzee to walk on all fours over the same distance. Perhaps this is because chimpanzees have anatomy that allows them to both travel terrestrially as well as climb trees for fruits, but humans are specialized solely for terrestrial movement. But when chimpanzees are trained to walk bipedally, they do not walk efficiently. They waddle, struggle for balance, and do not have a naturally smooth gait or arm swing. Presumably, this difficulty would be similar to that experienced by the first bipeds. So selection would have had to have been very strong in the early phases of bipedalism in order to overcome the difficulty of teetering on two legs. What could have driven selection to overcome the initial anatomical obstacles?

Thermoregulation, specifically the need to avoid overheating, could have been evolution's reason for bipedalism. This idea argues that it is better to walk upright in the stressful heat of the midday sun on the arid East African savannah than on all fours, because an upright posture means less skin is under the direct glare of the sun's rays. The big human brain is sensitive to heat stress, so such a strategy could have been beneficial. Plus, there are other adaptations dedicated to cooling the body, which indicate its importance, like the increase in sweat glands and the loss of body fur which aid in evaporative cooling of the skin.

Perhaps bipedal walking evolved from a vigilant upright stance used to watch for predators from the trees or while walking between tree patches. The drying of East Africa would have led to increasingly patchy woodlands and this hypothesis is linked to one of the oldest ones known as the "savannah hypothesis" that argues hominins had to move to the ground simply because there were fewer trees on the drying African savannah. However, it is clear now, based on paleoenvironmental reconstructions of the sites where *Sahelanthropus, Orrorin,* and *Ardipitheucus fossils* are found, that the earliest hominins, which presumably were somewhat bipedal, were living in forests.

Standing upright freed the hands of their locomotor role and Darwin, Owen Lovejoy, and others have emphasized the role of manipulation in their hypotheses for bipedal evolution. Chimpanzees and gorillas can carry objects in one hand, bracing it against their hip while walking in a "tripod" technique and they can also transport objects by wedging them between their mandible and their chest. But bipedal hominins would have had the use of both hands to make tools and carry them and selection was definitely strong for manipulation since by australopith times it is clear that their hands evolved precision grip capabilities that were better than chimpanzees. The thumb is longer and its joint is more mobile like a human's. Anatomical evidence for bipedalism predates the first stone tools on record by at least 2.5 million years, but the use of nonpreserving materials could have been just as significant (see the section on "Tool Use" in this chapter). Another advantage of free hands would for carrying food away from dangerous areas, like a carcass where scavengers are lurking. Carrying food to a potential mate in exchange for sex or to provision a family has also been offered as a hypothesis for bipedalism.

Bipedal walking could simply be an extension of standing bipedally to feed from small trees or a bipedal posture while feeding in the trees. Chimpanzees and gorillas stand bipedally to feed when they need to and their infants are, in fact, highly bipedal and clamor around on their mothers to keep their balance and, in turn, their mothers will sometimes walk bipedally to carry them.

No matter the cause and the driving force (it could be a combination of all of them to a degree over space and time), the freeing of the hands led to major specializations in the arm, wrist, and fingers for handling objects and the transfer of all the locomotor functions to the legs (aside from the use of the arms to swing by the side) led to major specializations in the entire body for walking and running on only two legs.

Exactly how the transition from arboreality to bipedality occurred (gradual or fast) is still unclear, but dietary and tool use hypotheses are the strongest. Once bipedalism evolved, however, all traces of arboreality did not disappear. Between 6 Mya and 2 Mya, hominins retained many features associated with arboreality like long arms, curved fingers and toes, and short legs. The question is: did the echoes of arboreality remain until 2 Mya because hominins were still arboreal to some degree or does adapting to bipedalism just take a long time?

Although the debate spans the Pliocene hominin record, the *Australopithecus afarensis* skeleton "Lucy" (AL 288-1) is usually the poster child for both sides of the issue, as is the rest of the species since the fossil record for it currently offers the most available evidence for a transitional hominin form.

Scientists in one camp which includes Owen Lovejoy say that Lucy was a biped just like modern humans. They point to the broad flared ilium (i.e., her basin-like pelvis), the "carrying angle" of the femur (the angle of the thigh bone at the knee), and the location of the foramen magnum under the skull. They argue that the Laetoli footprints (which were probably made by *A. afarensis*) clearly show a less diverged big toe, a well-developed arch (for storing energy and absorbing shock that is absent in flat-footed apes), and a clear depression for the impact of heel strike followed by toe-off (a uniquely human gait pattern) (refer back to Figure 3.6).

But those in the opposite camp, like Jack Stern and Randall Susman, argue that *A. afarensis* retained a heavy reliance on the trees for survival and were therefore skilled climbers. They indicate that Lucy and her species share similarities with climbing and suspensory adaptations in great apes like a funnel-shaped thorax, long curved phalanges on the hands and feet, a climber's shoulder (scapula) anatomy, relatively short legs, and relatively long arms.

With the discovery of more fossils it will be possible to determine whether australopiths remained at least part-time tree-dwellers throughout the Pliocene or not, but it is clear, even as the fossil record stands now, that fully habitual bipedalism, and its entire suite of humanlike locomotor adaptations, arrived by 1.8 Mya with *H. erectus*.

Aquatic Ape Hypothesis

Every so often the Aquatic Ape Hypothesis (AAH; also called Aquatic Ape Theory) washes up in the news media, but there has only been one address of AAH published in a peer-reviewed scientific journal. The idea that humanity sprung from aquatic past was first put forth by Alister Hardy (1896–1985), a Fellow of the Royal Society and a biological oceanographer. Then author Elaine Morgan took up the idea and has since written several books in support of Hardy's hypothesis.

Supporters of the AAH point out that hominin fossils are found in close proximity to water, and many times near large bodies of water, like the lakes Turkana, Tanganyika, and Victoria. They posit that an aquatic or at least a semi-aquatic ancestry may explain major physical differences between humans and other primates, like swimming abilities, the ability of newborn babies to swim and float, and the relative hairlessness of our bodies like many other aquatic mammals. In fact, the epitomes of humanity, bipedalism, and a large brain, are also explained by the AAH: in order to wade through water, hominins became better bipeds and then feeding on fish and shellfish provided the nutrition to develop big brains.

Although the AAH gets frequent press, it is considered a just-so story by most paleoanthropologists for many reasons. Wading and swimming are not unique to the human condition. If given the opportunity, in the wild or in captivity, many other primates wade (even bipedally) and swim. Lowland gorilla silverback males splash violently with their hands in the swamps in symbolic displays and others comfortably stand and sit to feed on swamp grasses. Japanese macaques are known to relax in natural hot springs. Clearly nonhuman primates are capable of developing hydrophilic behaviors that have not resulted in bipedality, hairlessness, calculus, or poetry. However, this line of evidence can also be used by the AAH supporters, since if the affinity for water is shared by many higher primates, it was probably present in early hominins and could have been a driving force in human evolution.

There certainly are a lot of hominin fossil sites found near water. However, fossils are most often preserved near water because they are rapidly buried in sediment and in most cases water is the transporter of that sediment. Fossil sites without water burial or at least some water interference are rare unless they are found in extreme desert or arctic conditions or in cases of instantaneous burial by volcanic lava or ash. Paleontologists are well aware of the preservation bias and expect fossils to be found near ancient rivers, streams, deltas, and lakes because of the preservation benefits they provide.

Although the AAH is not a strong or mainstream scientific hypothesis, we should not overlook the possibility that ancient hominins dipped, waded, and wallowed to stay cool like we and many other mammals do.

REDUCED BODY HAIR

Humans have dramatically reduced body hair even though it is important for protection against the climate and the sharp teeth and claws of other animals. Body fur can also be used to signal with color and patterns and also through piloerection, or raising the hairs to puff up and look big and aggressive.

Not only are human functionally "naked" but we are sweaty too. We have more sweat glands than any other primate and are some of the sweatiest mammals as well. It follows that the probable reason for our loss of body hair was due to the need for increased sweating. There is no fossil evidence for the evolution of human body hair so genetics and adaptations of other human traits guide the hypotheses.

Nina Jablonski argues that the need for thermoregulation while walking around with a big hot brain in the sunny hot periods of the day (presumably while foraging, hunting, and scavenging) led to the selection for body fur loss in early members of the genus *Homo* about 2 Mya. The loss of body fur was accompanied by the evolution of more sweat glands for copious sweating (and hence body cooling) and the evolution of darker skin for protection from harmful ultraviolet radiation (UVR). The combination of body fur loss, increased heat tolerance through sweating, and darker skin enabled hominins to travel further and spend longer periods of time under the hot sun. The *MCIR* gene associated with skin pigmentation points to a date of 1.7 Mya for the emergence of dark skin, which is consistent with the hypothesis that body fur loss and dark pigmentation evolved in concert around 2 Mya.

Hair is abundant on the top of the head where the sun hits the body directly and in the underarm and pubic regions, presumably for protection against chaffing and rubbing, or for storing scents that signal fertility and identity. All the hair on the body is actually only one of two types: terminal hair occurs on the head, eyebrows, and eyelashes but vellus hair is everywhere else. These types of hair differ in their lasting ability: they all grow at about a half an inch per month but they have different growing durations before they drop out of the skin.

Comparative genomics will someday illuminate the genetic history behind our unique body hair distribution. To investigate the loss of body fur we can look for candidate genes in the genomes of hairless mice, naked mole rats, hairless cats and dogs (with furry outgroups for comparison, of course) to see if humans have anything similar.

One candidate gene for long hair is known from mice (called "angora mice") with a mutation in *FGF5*, a gene that in its normal state functions to stop hair growth. Perhaps humans have a genetic mechanism for ignoring this hair-growth-halting gene as well. There are also several genes for keratin, the substance that makes up hair. Many of these genes are identical in humans, chimpanzees, and gorillas but one is different in humans. It has no apparent function in humans, but it codes for proteins in the other primates. The mutation in humans occurred about 250 Kya which could be the time when human head hair took its current ever-growing form. More studies on these genes will need to be done to understand this process better.

If body hair was reduced, why was head hair increased? It could be protection for the brain from the direct rays of the sun. It could be sexually selected for because of beauty. It could be a fitness indicator for health and social standing. Human hair needs extra grooming and taking the time to groom one's hair or to have someone else groom it could signal importance or status. The Venus figurine from Willendorf, Austria (Figure 5.4) had an elaborate hairdo, showing us an early glimpse at the importance of good hair by 23 Kya.

BODY SIZE, SHAPE, AND STRENGTH

Many human adaptations follow the same patterns as other mammals around the world, especially those that are determined by thermoregulatory rules. There are general mammalian-wide relationships between a body's surface area and volume (SA/V) that are controlled by climate. Bergman's Rule states that mammals in colder climates tend to have larger bodies than animals in warmer ones. A larger body size decreases the SA/V ratio and thus reduces heat loss. Based on a similar need for heat conservation, Allen's Rule states that mammals in colder climates tend to have shorter appendages than animals in warmer ones. The lengths of the distal segments, the hands, forearms, feet and shins, are shorter in human populations that live in colder environments at higher latitudes, like the Inuit (Eskimo) of the Arctic, compared to those populations who live in hot regions near the equator, like Nilotic peoples of Sudan, whose bodies are built to dissipate heat.

With the appearance of *H. erectus* in the early Pleistocene the increase in hominin body size is accompanied by linearity (i.e., lankiness) in body build to maintain the SA/V ratio appropriate for the hot, arid climate (Figure 5.2). The longer arms and legs and slender torso relative to body size allow greater cooling potential through sweating. The Nariokotome *H. erectus* boy was linear like present-day Nilotic populations, and maybe even more so.

Benefits of the linear build are lost in humid, forested, closed environments. Small-bodied australopiths, like pygmy human populations, were probably able to inhabit humid, closed environments as well as open, arid environments but *H. erectus* probably preferred open, arid ones. *Australopithecus garhi* shows a mosaic of australopith and *H. erectus* limb proportions with its long arms and long legs, so it represents an intermediate form between the two groups.

Neanderthals show just the opposite of the *H. erectus* body build. They had an extreme adaptation to cold temperatures, even compared to the most cold-adapted humans. With their relatively short arms and legs and thick barrel-chested torsos they are classic examples of Bergman's and Allen's Rules. Male Neanderthals averaged about 5 feet 6 inches tall and 180 lbs and females were about 5 feet 1 inch tall and 160 lbs. Although their stature was shorter, their bodies were broader, thicker, sturdier, and more compact than ours.

Although humans are taller and larger than many of our primate relatives, we are weak in comparison. Chimpanzees are on the order of four times stronger than humans and it is evident in their bones.

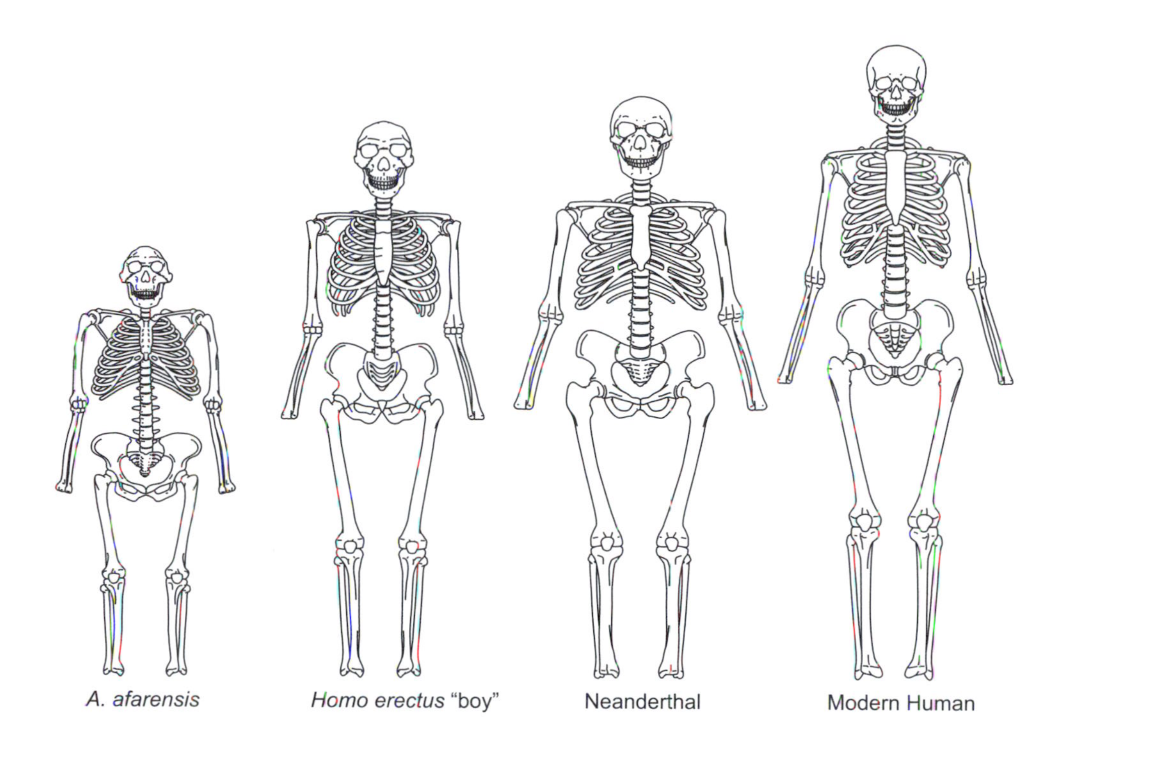

Figure 5.2 Skeletons of different hominins show their relative body sizes, bone thicknesses, and limb proportions. Bones of the hands and feet were omitted for simplicity and also because few were preserved with the (Nariokotome) *H. erectus* skeleton. *Illustration by Jeff Dixon.*

The skeletal infrastructure of a strong body is different from that of a weak one. Bones that anchor strong and big muscles show larger muscle attachment sites, are thicker and heavier, may have some curvature to them to resist forces better, and also have bony buttressing and struts inside them which are visible by x-ray.

Bones are constantly remodeling in response to the forces generated during physical activities (a principle known as "Wolff's Law"), which is why exercise is required to strengthen bones in order to avoid harmful levels of bone loss later in life. Osteoblasts are progenitor cells that form bone and osteoclasts are their resorbing counterparts. The thick outer layer of bone (cortical bone) reacts to stress by laying down more bone where it experiences forces, and resorbing bone where it is not under stress. For example, the humerus in the racket-wielding arm of elite tennis players, especially those who started playing as children, is much thicker than the other arm. Underused bones will resorb and atrophy instead of remodel with new bone like a healthy bone. Astronauts' bones are severely traumatized after spending lengths of time in zero gravity aboard the space station and scientists are inventing comfortable ways to tether the astronauts to the treadmill to add forces to their bones while they exercise.

Strength indicators from bones can also be tracked in the hominin fossil record and it is clear that fossil hominins were much stronger and more active than modern humans. Especially by *H. erectus* times, when body sizes and proportions were similar to ours, it is apparent that bones were much thicker and muscle attachments sites were much bigger as well. Neanderthals are an extreme example of skeletal strength with their thick heavy and sometimes curved long bones. But even before some modern human populations became sedentary with agriculture, their bodies were relatively weaker than those of previous hominins. Cultural innovations probably compensated for the need for *H. sapiens* to maintain physical strength.

TEETH

Because tooth enamel is made of sturdy material (hydroxyapatite), teeth make good candidates for fossilization. As a consequence, teeth are the most common part of hominin skeletons that are discovered and collected.

Teeth are distinguishable by species because their shape and size are linked to body size, diet (for shearing, cutting, crushing, grinding) and social behavior (long, sharp canines for mate competition). Differences in the number, size, shape, cusp patterning, and placement of the teeth provide a means of identifying ancestral and related species, and tracking dietary evolution.

If you smile at a mirror it is easy to see that there are different types of teeth in the human mouth. The cutting incisors in the front are like flat spades, the canines resemble small fangs or extra incisors, the premolars (what some dentists calls "bicuspids"), and the molars are bumpy, multipurpose grinding implements. This condition of having a diverse dental toolkit is called heterodonty. The extreme opposite condition, called homodonty, is seen in the jaws of a crocodile where all the teeth are sharp cones used for gripping and tearing.

All primates are heterodontic, but they can be distinguished from one another by the particular dental patterns they display, and the specific shapes and sizes of their teeth.

Humans have generalized dentition indicating omnivory like suids (pigs and their ancestors). In fact it sometimes takes a well-trained paleontologist to distinguish between a fossil suid and a fossil hominin molar if it has been broken or worn down through use in life.

During hominin evolution the canine teeth became increasingly smaller. Overall, humans have small teeth for the size of their head and body and have very little dental sexual dimorphism compared to apes and monkeys. Early hominins get smaller and smaller canines before evidence for meat-eating shows up (around 2 Mya) and this trend indicates that social changes were probably the prime mover for canine reduction, not dietary ones. As canine size lost traction in the mate competition game, either (1) mate competition decreased like in gibbons which form monogamous pair bonds, or (2) other forms of mate competition took hold that had little to do with tooth weaponry.

The biomechanics of teeth and the surrounding masticatory (chewing) apparatus are usually correlated with the types of food being processed. *Paranthropus* had large, flat, thick-enameled molars and an accompanying suite of strong masticatory adaptations in the bones of the skull (see Chapter 3). The teeth were clearly not built for shearing grass and seem to be adapted to handle tough plant foods such as hard seeds and nuts.

Individuals experience microscopic trauma to their teeth from food and these scars caused by food can be used to infer diet of fossil hominins. The pits and scratches on teeth caused by eating different types of foods can be linked to certain types of food based on size, intensity, frequency, and location (e.g., incisors versus molars). The first studies of microwear in the 1970s were performed with a scanning electron microscope (SEM), but thirty years later scanning laser microscopes are the preferred tools. *Paranthropus* teeth show microwear from hard, but not abrasive food, that was determined to be nuts or seeds with hard cases, supporting the nickname "Nutcracker Man" that Louis Leakey gave to the OH 5 *Paranthropus boisei* skull.

Wisdom Teeth

The third molars are called "wisdom teeth" because they erupt at the onset of adulthood between eighteen and twenty-two years of age. Many people do not have large enough jaws to accommodate their third molars and they can become impacted, or stuck inside the jaw when it is time for them to erupt. People with access to dental care will often get their third molars removed to avoid pain, infection, and crowding of the teeth, which leads to malocclusion, more pain, and difficulty in chewing.

It has long been assumed that such problems are the result of finely processed soft foods overtaking the diets of many agriculture-based societies beginning about 10,000 years ago. Eating soft foods for an extended period during childhood does not stimulate the jaw to grow as large as in those who eat a tough or coarse diet, which causes bone to grow and to buttress itself against the forces.

However, one fossil skeleton from a site in France contradicts the agriculture theory for the wisdom teeth problem. The 15 Kya "Magdalenian Girl" from the Cap Blanc rock shelter had impacted wisdom teeth, but they went unnoticed (which is why she was called a "girl") until powerful x-ray imaging allowed scientists to see them inside their crypts in her jaw. Now the specimen has been aged to between twenty-five and thirty-five years at the time of her death and she should now be called the "Magdalenian Woman." Since she predates the earliest evidence for agriculture, soft processed foods were not necessarily to blame for her problem. However, it is possible that pre-agriculture humans were processing wild foods similar to the way later humans would soon start to process domesticated foods. Further archaeological evidence will shed light on this in the future.

TOOL USE

Humans are not the only animals to use tools. Sea otters, naked mole rats, and owls are just some of the other tool-users on the planet. Dolphin mothers teach their daughters to use sponges to protect their snouts from stinging creatures while foraging on the sea bottom. Within the primates, habitual tool use (like the use of hammers, probes, and scrapers made of rocks or vegetation) is mostly limited to orangutans, chimpanzees, bonobos, gorillas, and capuchin monkeys (New World monkeys). Gorillas and chimpanzees (especially from the Tai Forest in Côte D'Ivoire) crack open oil palm nuts by smashing them between two rocks with a "hammer and anvil" technique. Gorillas have been observed to use sticks and branches as crutches to cross water, and chimpanzees will use leaves as sponges to sop up drinking water.

There is a distinction, however, between using a found object as a tool and modifying an object for use as a tool, something very few animals do, aside from chimpanzees and humans. Chimpanzee females and their daughters will prepare a stick and then dip it into a termite mound to extract the nutritious insects. The brush-like tip of a termite fishing probe is made by pulling a plant stem through clenched molars, straightening it to a point by wetting it with saliva, and then carefully manipulating the fibers by hand or mouth. Sticks and branches are also used to kill much larger prey. Chimpanzees have been observed to use prepared branches as spears to kill bush babies (see the section on "Scavenging and Hunting" in this chapter).

Certainly early hominins were using plant materials as tools, but these artifacts do not preserve like stone tools do. Because chimpanzees do not manufacture stone tools in the wild—that is, they do not strike off bits with the intent to use the newly modified rocks—stone tool artifacts are attributed to the handicraft of our hominin ancestors. Even when capuchins and chimpanzees are taught to make stone tools in captivity, they fail to produce even the simplest Oldowan-like tools.

Cognitive power determines the ability to make stone tools but so does anatomy of the wrist and hand. There are a few hand and wrist bones of australopiths and they show derived features like a flexible wrist, a long thumb, and broad fingertips that are related to humanlike manipulation and the capability to produce stone tools. Surprisingly, evidence dealing with the enlarged *gluteus maximus* muscle in humans also contributes to our understanding of hominin tool use. Mary Marzke and her colleagues pointed out that an *Australopithecus* ilium from South Africa shows that the *gluteus maximus* muscle that attached to it in life was large and similar to that in humans. Relative to that in chimpanzees, the human *gluteus maximus* arrangement provides a mechanical advantage during tool use. Our *gluteus maximus* controls the rotation and movements of our trunk which keeps us balanced and stable when we are using tools, throwing objects, clubbing, digging, and basically doing anything that involves vigorous arm movements. Thus, it appears from the hands and the buttocks that, by 3 Mya, australopiths had the body for serious and habitual tool use. But again, bipedalism and tool use must be linked, because the *gluteus maximus* muscle is also important for running. Since

Figure 5.3 Acheulean hand axes and other tools literally litter the landscape at the early Pleistocene site of Olorgesailie located in the Great Rift Valley of Kenya. *Photograph by Holly Dunsworth.*

running (as opposed to walking) is a series of one-legged jumps, the gluteal muscles are highly engaged in the activity.

Stone tools are found in association with evidence for hominin meat-eating behavior. By the early Pleistocene there are sites like Olorgesailie, Kenya, that show how important stone tools had become. There are hundreds of hand axes and other stone tools scattered across the ground and associated with *H. erectus* remains nearby (Figure 5.3). The ubiquity of the hand axes and their consistency in shape in space and time (over 1 million years!) begs the question of their function. Certainly they served, as their name implies, as handheld axes for skinning and butchering animals. Plus, the flakes that came off of them as they were made would have been useful for cutting implements as well. But, they also would have been useful for digging up tubers or perhaps for digging for underground water sources. One hypothesis calls on their aerodynamic nature to postulate that they were launched at prey as "killer frisbees." It is easy to imagine such a scenario from the looks of Olorgesailie.

DIET

Paleoanthropologists are concerned with diet evolution because it is so strongly correlated to other variables (like brain size, body size, intelligence, activity levels, geographic range, tooth shape, skull shape, etc.) and, not insignificantly so, because food intake is the primary means of survival. Diet has also changed considerably throughout human evolution. As we have already discussed, tools and teeth lend clues to hominin diet. However, there are several other useful methods for reconstructing hominin diet (others are discussed in the section on Scavenging and Hunting in this chapter).

There are ecological and energetic rules involved in diet. Energy from the sun is incorporated into plant matter, which is then incorporated into the flesh of herbivores and is finally taken up by carnivores. At each level of the trophic pyramid (i.e., the ordering of groups of animals according to diet, with carnivores on top of the pyramid) there is a decrease in energy available per unit of biomass. As a general rule, primates with large body sizes tend to eat vegetation because of its abundance relative to nectar and insects, which are sufficient for nourishing small-bodied species.

We, literally, are what we eat. Trace elements in fossil teeth and bones tell of an individual's trophic level. For instance, the metal strontium (Sr), which is usually studied in relation to calcium as the ratio Sr/Ca, is taken up by plants through ground water but is only taken up by animals that eat the plants. Low levels of Sr indicate carnivory and high levels indicate herbivory. The same type of trophic information can be gleaned from levels of barium, magnesium, and zinc as well. The hominins at Swartkrans have omnivorous (intermediate) Sr levels, which contradicts a common assumption that australopiths and *Paranthropus* were purely herbivorous.

Stable isotopes from teeth and bones also tell a diet story. Carbon isotopes are particularly informative and are based on the characteristics of the two different paths of carbon dioxide fixation in photosynthesis. The C3 photosynthetic pathway differs chemically from the C4 pathway, so C3 plants have detectably different ratios of carbon isotopes than C4 plants. C3 plants are mostly trees, shrubs, herbs, and temperate grasses. C4 plants include maize, some millets, sorghum, and warm grasses. Animals that eat these different plants show the different ratios. For example, giraffes are browsers that eat mostly C3 plants and zebras are grazers that eat mostly C4 plants. Therefore, fossils of individuals that have higher carbon isotope ratios (C4) are labeled grazers and those that have low ratios (C3) are browsers or they are eaters of grazers or eaters of browsers, respectively.

Paranthropus shows a mixed carbon signature like aardvarks that eat primarily termites (which fall under the C4 signature). So *Paranthropus*, like its chimpanzee relatives, could have incorporated termites into its average diet. Analysis of bone tools from South African cave sites supports this notion. The tools are remarkably similar to experimentally made termite fishing wands with polished, rounded tips and parallel microscopic striations from the soil. Carbon isotopes also indicate *Paranthropus* ate tropical grasses and sedges, woody fruits, shrubs, and herbs. Differing signatures that alternate in the teeth indicate their diet was seasonal and switched between diverse plants depending on the time of year.

Of course, direct evidence of diet can be found in the contents of an individual's stomach as well as in fossilized dung (coprolites) and vomit. Seeds in the hominin coprolites from Terra Amata (380 Kya) suggest that the camp site was seasonal since the particular seeds were only available during the summer. Unfortunately animals do not normally fossilize with preserved organs. If fossil animals did preserve stomachs, the contents of the stomachs would be direct evidence for diet, or at least for the last supper. The famous bog bodies of Great Britain and Denmark often had full meals preserved in their stomachs.

Food-related diseases can tell volumes about an individual's diet as well. Such is the case of KNM-ER 1808, an adult female *H. erectus's* partial skeleton from Koobi Fora, Kenya. Her long bones had a covering of woven bone, which is a sign of disease. Bone should be smooth, not bumpy as hers were and the pathology is consistent with modern cases of hypervitaminosis A, a disease caused in modern humans by ingesting an overdose of vitamin A. But, in the early Pleistocene, the disease was likely incurred from eating too many carnivore livers, as carnivore livers are high in vitamin A relative to anything else available on the East African savannah. If a *H. erectus* was able to overdose on an animal product around 1.5 Mya, it is clear that they were skilled at obtaining meat by then.

Tapeworms

The human tapeworm offers an intriguing angle on the evolution of meat-eating. Domestic animals have long been blamed for giving humans tapeworms (*Taenia*). So when genetic analyses were performed on tapeworms from a variety of animal hosts, scientists expected to find an origin for the human tapeworm around the time that agriculture first became prevalent around 10,000 years ago. On the contrary however, the results showed that human-specific tapeworms are not descendents of tapeworms living in modern domestic animal hosts. Human tapeworms evolved further back with the shift to meat-eating behavior when humans joined the carnivore guild.

The tapeworm life cycle involves two different hosts. First it lays eggs on the ground that are eaten by a grazer (like an antelope) which is the tapeworm's "intermediate host." The egg forms an embryo inside the muscle of the animal. When a carnivore, or "definitive host," eats the intermediate host it ingests the young tapeworm that grows into an adult tapeworm and leaves the carnivore with the feces. The tapeworm, then, has been transported back to the ground where it will lay eggs and continue the life cycle.

Tapeworm gene sequences are most similar between humans and hyenas and big cats in Africa, so hominins probably picked up tapeworms from eating the intermediate hosts of tapeworms specific to those carnivores of the prehistoric African savannah. Molecular clocks of the sequences point to an origin around 1.71–0.78 Mya, which is during *H. erectus* times.

SCAVENGING AND HUNTING

Humans share a common herbivorous, or plant-eating, ancestor with living apes and early hominins were mostly vegetarians who ate fruit, nuts, tubers (roots), and also ate insects like termites. Like other herbivorous mammals, including monkeys and apes, humans cannot synthesize vitamin C—an unnecessary skill with a diet comprised of vitamin C-rich vegetable matter. But the drastic shift in hominin brain and body size around 2 Mya is linked to a shift in diet toward carnivory. Scavenging and hunting during the hot daytime hours on the East African savannah is a significant component of the evolution of the suite of characters that differentiated early *Homo* from australopiths and *Paranthropus* and eventually led to humans.

Although meat-eating holds an important place in our evolutionary history, it did not emerge uniquely in our lineage. Mostly fruit-eating chimpanzees at Gombe, Tanzania, are known for their habit of killing and eating red colobus monkeys to the point of significantly reducing monkey populations in the region. They have even been known to kill small bovids and wild pigs. Chimpanzees also use tools to hunt for meat. Jill Pruetz and Paco Bertolani observed male, female, and even juvenile chimpanzees in Senegal fashioning spears from branches (similar to the way females fashion termite fishing wands) and using them to kill bush babies (nocturnal strepsirhines that sleep during the day). They jabbed the spears into hollow trees where bush babies were nesting in order to immobilize them before breaking into the tree to catch and eat them.

It is hard to imagine that our hominin ancestors lacked such hunting sophistication but unfortunately spears like those made by chimpanzees do not preserve in the fossil record for us to find out. In order to track the evolution of human meat-eating by scavenging and hunting, we must follow the record of stone tools and devoured animal bones. Stone tools are found as early as 2.6 Mya and, soon after that, around 2.5 Mya, the earliest cut-marked bovid fossils come from Bouri, Ethiopia. Although these dates are normally cited as the earliest considerable incorporation of meat into the hominin diet, neither site has both cut-marked bones and stone tools.

The beds at Olduvai Gorge (dating to as early as about 2.5 Mya) provide direct association between tools and butchered bones. Pat Shipman and Rick Potts studied the cut marks and the carnivore tooth marks on the Olduvai faunal assemblage and suggested that hominins were scavenging meat. Cut marks on bones indicate butchery and with the aid of a microscope they can be distinguished from carnivore tooth marks. A stone flake, no matter how fine-grained the material is, will never have a perfect straight edge, so that when it is dragged across a bone during use, its jagged edge leaves many fine parallel lines. On the contrary, carnivore teeth leave a singular, rounded groove on bones.

At Olduvai, cut marks were found superimposed on tooth marks and vice versa implying that either hominins were the primary hunters/scavengers and carnivores were secondarily scavenging from kill sites, or that hominins were scavenging from carnivore kills. The second scenario, with hominins as scavengers, is usually the case at the earliest sites. Marrow inside the bones is especially nutritious and scars from bashing open bones with stone tools are also distinguishable from the marks from a carnivore's bone crushing teeth.

In the absence of carnivore tooth marks for hunting versus scavenging evidence, there are other means of establishing meat-eating. Stone tool cut marks located at areas of skeletal articulation are signs of butchery (e.g., at the joints where ligaments and tendons are connecting the bones). Depending upon whether the archaeological site is a kill site (where the animal is butchered and carried "home") or a living site (or "living floor"), the absence or presence of the bones associated with the meaty and delicious parts of animals also indicate whether or not hominins had first dibs on the carcass or whether they scavenged after a carnivore. Even if there are no cut marks it is sometimes possible to identify a fossilized carnivore kill because each carnivore species has a particular way of devouring an animal. Some eat the feet and ribs and leave the rest of the bones, but some eat everything, even the vertebrae, and leave only the skull and teeth. Bob Brain was one of the first to document the different eating preferences of carnivores because he actually fed them baboon carcasses and then analyzed the parts that were left behind.

Selection must have been very high on hominins who obtained meat since scavenging from a carnivore kill would have been extremely dangerous. Hominins could have developed sly methods of creeping toward a carcass, but also could have coordinated groups to chase away the carnivores from their food. After acquiring the skills to chase away a carnivore from a kill, it hardly seems like a big behavioral leap for hominins to hunt and kill their own prey. Because chimpanzees do not obtain meat through scavenging we would not necessarily assume hominins did that either if it were not for the abundant evidence from sites like Olduvai Gorge and the overlapping of cut marks and tooth marks.

The evolution of meat-eating and then predatory behavior is probably the root of many modern sports and athletic endeavors. Endurance running and overarm throwing no doubt helped early hominins hone their hunting skills. Genes like *ACE*, although it does not exist in all populations, help explain how humans can work harder with less fuel and increase their endurance. Throwing would have been of great importance since it allows action-at-a-distance. And William Calvin has even suggested that because of the complex neural control involved in accurate overarm throwing, it is linked evolutionarily to brain size increase and language.

Meat intake could was definitely beneficial and possibly necessary for growing a large brain. Leslie Aiello postulated that to balance out the energetic cost of growing and maintaining a big brain, hominins evolved shorter guts (i.e., selection favored hominins with shorter intestinal segments). Humans have relatively smaller guts for their body size compared to apes, which is a shared feature with carnivores. But intestinal material is nearly as energetically costly as the brain and perhaps the body compromised for the development of more "expensive tissue" in the brain and less in the gut, thus reinforcing a shift in diet.

Meat-eating and related strategies separated *Paranthropus* from the *Homo* lineage. While *Paranthropus* was eating plant materials and termites, early *Homo* had shifted their diet to include meat and bone marrow and this is perhaps indicative of their sympatric speciation. Around the time that we see an abundance of archaeological data for meat-eating, there is a transition in hominin body size and brain size. A growing brain needs more energy for maintenance and the incorporation of meat into the diet may have selected for bigger brains and increased intelligence (needed for hunting skills) while feeding them. Several predictions can be made about a species that becomes a predator. Based on knowledge from our current fossil record, *H. erectus* fits many of these predictions. Namely *H. erectus* shows an increase in body size, an increase in geographic range, a major technological shift (from Oldowan to Acheulean tools), and an increase in sociality. The last point is debatable because evidence is indirect, but some would point to the evolution of group hunting as support.

FIRE

The manipulation of fire led to dramatic changes in hominin behavior. Fire offers warmth and would have enabled hominins to extend their habitat to colder regions (either at higher altitudes or higher latitudes). Fire provides light at night or in caves; it can be used as protection against predators; it can be used in hunting to control animal movements; it is useful for making tools out of wood and some types of stones that are best when hardened by heat; it also offers a way to detoxify and soften foods (both plant and animal) through cooking.

Archaeological evidence for fire includes ash accumulations and burnt soil, rocks, bones, wood, charcoal, or other artifacts. Discoloration and chemical and structural changes of these materials are indicative of burning and the atomic changes that occur after superheating objects can be traced using thermoluminescence.

The earliest evidence for controlled fire is found at Gesher Benot Ya'aqov, a site in Israel that dates to 790 Kya. It has Acheulean tools, burned flint artifacts, charcoal fragments, burned wood, fruits, and grains. The *H. erectus* site of Zhoukoudian, China, also has a preserved hearth dating to about 500 Kya as does the *H. erectus*/Archaic kill site of Terra Amata, France, at about 380 Kya.

Short-term campfire signatures (bowl-shaped soils) have been discovered at much earlier sites, like Koobi Fora (1.6 Mya) and Chesowanja (1.4 Mya), Kenya. There are also some burnt bones at Swartkrans cave in South Africa from about 1.5 Mya. But these more ancient sites have not yet been confirmed as sites with controlled fire.

Once it is established that there is indeed evidence for fire, it is difficult to distinguish between its natural existence and human control of it. High temperatures caused by concentration of a fire at a hearth exceed those for natural fires that sweep across a landscape. Then whether or not humans opportunistically controlled fire (taken from a naturally occurring source) or they sparked the fire intentionally has so far proven impossible to tell. The latter implies a level of sophistication that is "human" while the first does not. Although, the opportunistic use of fire is a large step ahead of apes, no matter how it was obtained. Controlled fire and the use of hearths at home bases and living sites clearly became popular by the time Archaics were the focal point of hominin evolution.

REPRODUCTION

To some, like Owen Lovejoy, the truly distinct human traits are involved in reproduction. Females are receptive to mating even when they are not ovulating, couples mate privately, and they often mate only with one partner for a long stretch of time and even for life. This is a striking contrast to reproductive behaviors of baboons and chimpanzees, for instance, which share many similarities with humans in biology, sociality, and intelligence.

Female baboons and chimpanzees usually only mate when they are either in estrous, which is a segment of their ovulatory cycle when their anogenital perineum engorges into a bright pink swelling to signal the arrival of their fertile time. Often an abundance of food will trigger the females of a group to come into estrous and males reach heightened levels of competition around these females. Human females have conceived ovulation, which is hidden from males as well as from themselves. This situation opens them up for perpetual mating opportunities and may require males to find alternative ways to detect female ovulation, like, for instance, other visual or even olfactory (scent) cues. Additionally, concealed ovulation has the potential to place males in a state of perpetual competition.

Of course, there is much less privacy in the baboon and chimpanzee worlds, but they may mate secretly if codes of status and hierarchy are being breached. Also, for most primates, it is more often better to mate with more than one partner, not just in one mating cycle but throughout life. Most humans use the monogamous strategy, that is they form enduring pair-bonds (either serially or for life), but many humans are polygamous, forming long-lasting relationships between one male and multiple females instead. In either case, there is much less human male competition and also an unusually high level of male parental investment. Because of their extended period of growth and maturation, mostly dealing with the incredible amount of time required to grow the large human brain, human infants are altricial, which means that they are vulnerable and highly dependent on their parents after they are born. The opposite condition is to be precocious like newborn horses that can take off running soon after they touch the ground because they complete more growth in the womb.

Humans clearly have their own reproductive strategy that stands apart from even their closest relatives and we have the unique anatomy to show for it. *H. sapiens* show low levels of sexual dimorphism in body size, strength, physiology and are the least dimorphic out of all the great apes, but we are not as monomorphic as monogamous gibbons. Human males are larger and stronger and have higher metabolic rates than females. They have higher juvenile mortality rates and they attain sexual maturity later than females. Hair and fat distribution and abundance differentiate the sexes and so do the somewhat large and conspicuous sex organs. Humans fall between chimpanzees (large) and gorillas (small) in testes size for body weight. In other words, humans have somewhere between highly competitive sized (chimpanzees) and essentially competition-free sized (gorillas) testes. Males in a primate species where females mate regularly with more than one male tend to have elaborate penises that are either highly stimulatory or bristle-like for direct sperm competition inside the vagina. Those species where females tend to mate with only one male during an estrous cycle have much less elaborate anatomy and resemble the human form. However, the human penis is much larger and lacks a baculum (penis bone) which some suggest is the result of sexual selection by females, especially since bipedalism made the male genitals much more conspicuous.

Bipedalism may have helped shape the sexual anatomy of human females as well. The breasts are larger than expected when not lactating and the distribution of fat on the hips, thighs, and buttocks lowers the waist-hip ratio (WHR). Both are probably fertility signals and fitness indicators. A WHR of 0.7 is nearly cross culturally considered the "ideal" female figure, whether the woman is thin or heavy. Women with that classic hourglass silhouette have optimal estrogen levels that is correlated to fertility. So although no males are literally measuring and calculating the WHR of potential mates, they are able to pick up on the cues from a woman's body that are linked to evolutionary fitness.

The australopiths had a high degree of sexual dimorphism in body and canine size, which was greater than that in humans so the human reproductive strategy had not evolved in australopiths. Early hominins were probably polygynous in their mating behaviors like modern chimpanzees but by the time *H. erectus* arrived the strategies changed. Fossil evidence indicates that sexual dimorphism greatly decreased by *H. erectus*. Plus, the species grew in body size but retained a small hip-breadth (at par with australopiths) for reasons to do with bipedal efficiency as well as climate adaptations in body proportions.

An increase in brain size must accompany an increase in body size, so *H. erectus* could not have fit the same sized infant's brain and skull as expected for its body size through its narrow hips. That is, *H. erectus* may have shortened gestation (i.e., the period of fetal development in the uterus) to be physically capable of giving birth to larger brained babies through its relatively small birth canal. An earlier birth results in a more helpless, less developed, altricial infant. So it is probable, that with *H. erectus*, higher levels of parental investment, especially from the father (paternal investment), began to evolve. Selection would favor such a change if it fostered brain growth and development, especially if selection was acting strongly on brain size increase in the species. Other major changes are correlated, like the incorporation of meat into the diet, food-sharing, and the creation of home bases where males provisioned females and offspring and where females localized their large contribution to family diet from foraging and gathering (sexual division of labor). Once heavy paternal investment in offspring and the formation of pair-bonds became successful adaptations, females would have also chosen to mate and bond with males that could not only provide food and protection but that possessed dependable and fatherly attributes like kindness, generosity, and trustworthiness. Less time is spent foraging when the quality of the food goes up, so with the addition of meat and other high-quality foods into the diet, presumably more free time was spent at the home base. When this occurred, intellectual qualities like elaborate language, singing, music, wit, and even dancing would have become largely important in mate selection. There is no evidence that language evolved as early as *H. erectus* times, but singing, dancing, and humor cannot be ruled out in the early Pleistocene.

Menopause

Menopause, or the termination of reproduction, is an evolutionary puzzle. How could natural selection favor it when it clearly prefers increased reproduction? Human females experience menopause around age fifty when their finite supply of eggs is exhausted and reproductive processes slow and eventually cease.

Some, including Kristen Hawkes and her colleagues, posit that menopause could be an adaptive trait. Females that stop reproducing at menopause, start helping their reproductively active daughters provide for their grandchildren. While her daughter is lactating (an energetically expensive process), the grandmother helps subsidize the diet of the non-nursing children by gathering additional food. In this way, grandmothers help their daughters have more children and also increase the children's survival rates. Presumably, a grandmother's genetic material, which includes the genes for living long past menopause, is passed onto her better surviving children and grandchildren.

Another explanation for menopause, put forth by James Wood and colleagues, does not consider it an adaptation, despite the benefits of grandmother care. According to Wood's hypothesis, menopause and post-menopausal life are not themselves beneficial—rather they evolved because of antagonistic pleiotropy, a process whereby genes that have harmful effects later in life can be actively selected for if they have beneficial effects earlier in life.

In young women, a process known as follicular atresia helps to maximize fertility, but also causes the reproductive system to speed through the finite supply of eggs. Menopause is therefore a compromise, not an adaptation, because women give up the ability to reproduce later in life in order to experience high levels of fertility earlier in life.

Perhaps we should not ask why menopause evolved, but instead investigate when women started living beyond their reproductive years and why females, in comparison to males, have limited gamete production in the first place.

LANGUAGE

Language is such an integral part of being human that it is not even learned. Instead, language develops naturally like an organ or what Steven Pinker calls the "language instinct." By the age of five years, humans know all the rules of language. Noam Chomsky named our automatic propensity for learning and mastering the complex and sometimes illogical rules of grammar, our "language acquisition device," and his theory was strengthened by his discovery of a "universal grammar" which is the common basis for all human languages.

It is very difficult to trace direct evidence for human speech and language evolution in the fossil record. None of the soft anatomical parts involved in speech—like the tongue, larynx (voice box), and soft palate—are preserved and the bony parts that do fossilize do not reveal very much useful anatomy.

There is still no conclusive evidence that any hominin other than modern humans had language or could even speak like us. The small, floating horseshoe-shaped bone in the neck called the hyoid was thought to hold some clues, and since it is preserved in the Kebara Neanderthal (from 60

Kya in Israel), hopes were high. But the anatomy of the bone, although humanlike and not at all chimpanzee-like, was found to be identical to a pig's, so it is not diagnostic for language abilities. Although, when the length of the neck and the base of the face and jaw are considered, the Neanderthal vocal tract appears to have ape-like proportions. Apes and human babies have a high larynx and they can suckle and breathe at the same time. But as humans develop past infancy, their larynx lowers and this is what helps us make a wide range of vowel sounds in the throat, but inhibits us from breathing and eating at once.

Because specific areas of the brain are linked to speech and language, there may be potential to glean evidence for language from brain endocasts. For instance, that of a particular *H. erectus* cranium SM 3 (from Sambungmacan, Indonesia) has a pronounced Broca's cap, a bump on the left frontal lobe corresponding to Broca's Area, which is the anatomical region associated with speech production and language processing. But, Broca's caps can be found on chimpanzee endocasts so they are not foolproof indicators of language. Furthermore, the brain is wrapped in a covering of meninges, which is composed of the layers called pia mater, arachnoid, and dura mater, that dulls the brain's impression for an endocast, making interpreting the bumps on the brain an even more difficult task.

Complex coordination and control of the muscles of the trunk are necessary to control finely tuned breathing patterns during human speech. So it is possible to link the size of the spinal cord to the quantity of nerves required to monitor breathing and speech. The size of the hole through the vertebrae, the vertebral foramen, is small in *H. erectus* (based on the Nariokotome Boy), implying that his spinal cord was small too. Thus, there was probably not as much enervation of the thorax in *H. erectus* as in modern humans and it is unlikely that the Nariokotome boy could speak like us. But that does not rule out the possibility that he had a rudimentary language.

Because the anatomical evidence is inconclusive, we cannot rule out any particular hominin from having language. However, scientists have traditionally correlated symbols, art, and human culture and technology with the presence of language and none of that appears until after 100 Kya. However, cooperative hunting, with evidence of driving herds off cliffs by the middle Pleistocene, may be an earlier innovation that can be linked to language. The social complexity (and thus cognitive functioning, which is required for language) involved in cooperative hunting exists on a much larger scale than that accomplished by dogs and cats that hunt in groups.

The genetics of speech and language are beginning to unfold. Mutations in the gene *FOXP2* cause a person to struggle with motor control of the mouth and facial muscles so that word pronunciation is difficult. Persons with mutations at the gene also have deficiencies in certain aspects of grammar and cognition. Once *FOXP2* was identified, scientists searched for its function in normal humans and other animals. *FOXP2* is a regulatory gene, shared in similar forms by all vertebrates, that manages the activities of other genes, some of which are involved in language but others are not involved in language at all (which complicates hypotheses for its selection based solely on language). The human version of the gene differs from the mouse by only three amino acids and from the chimpanzee by only two, but it is possible that these two changes were of functional significance to the origin of language. Molecular clocks of human *FOXP2* point to a very recent origin within the past 200,000, but because it is a regulator gene it is highly unlikely that it would be the only gene involved in evolving language abilities. *FOXP2* reminds us that evolution is "descent with modification" as Darwin said, because something so crucial to making humans "human" probably evolved from primitive genes that are not unique at all to humans.

Selection must have acted very strongly on language the way it acted on bipedalism. Hypotheses for its earliest advantages include the ability to hunt more effectively by exchanging information about the physical and ecological environment. Robin Dunbar suggests that language evolved in order to exchange information about the social environment, or basically to gossip. Social communication is no trivial matter, since information about who can be trusted—that is, who plays the reciprocity game fairly—can be shared across a community (see the section on "Altruism and the Human Colony" in this chapter). With language, an individual would not need to keep track of every single one of the complex social and political relationships and networks in their community. By gossiping, that information could be shared and remembered much more easily. Language in this sense allowed humans to live communally in large groups and ramped up the importance of an individual's reputation if others could make information about him or her public knowledge. Language could also have been, as Geoffrey Miller suggests, a seduction tool favored by sexual selection.

Primates rely heavily on both vocal and nonvocal language and foundations for complex human communication exist in primates. As a whole, the group uses its increased numbers of facial muscles to convey emotions. White-handed gibbons use specific songs for warning others that a predator is threateningly close. Chimpanzees are aware of the

Table 5.2 When Are Hominins Humans?

"Hominin"	6 Mya	*Sahelanthropus*
Teeth	5 Mya	*Ardipithecus* or earlier
Bipedalism	5 Mya	*Ardipithecus* or earlier
Small canines	3.2 Mya	*Australopithecus afarensis*
Leg proportions	2.5 Mya	*Australopithecus garhi*
Stone tools	2.5 Mya	*Australopithecus/Homo?*
Meat	2.5 Mya	*Australopithecus/Homo?*
Geographic dispersal	1.8 Mya	*Homo erectus*
Body proportions	1.8 Mya	*Homo erectus*
Fire	800 Kya	*Homo erectus*/Archaics?
Brain size	500 Kya	Archaics
Symbols	100 Kya	*Homo sapiens*
Language	100 Kya?	*Homo sapiens*

importance of facial communication and have been observed to literally wipe fear smiles off their own faces in the presence of an aggressive group member in order to appear unafraid. Captive chimpanzees will naturally develop specific calls for specific food items like bananas and wild chimpanzees use specific vocalizations for snakes and threatening strangers, and they also have hunting calls.

HUMAN REVOLUTION

Evidence for what we consider modern culture and behavior—like adorning oneself with clothing and jewelry, making art, and performing rituals—does not actually appear when the first anatomically modern humans emerge around 200 Kya. Biology and behavior are not directly linked in hominin evolution, so evidence for human behavior does not appear until well after modern human anatomy evolved. Although the evolution of human culture and behavior happened gradually, the phenomenon is referred to as the Upper Paleolithic revolution. It is clear that hominins are humans by the time they leave advanced cultural debris behind, but it is hypothetically possible to extend humanness further back in time since human traits accumulated over millions of years (Table 5.2).

Culture is not unique to humans but it certainly is exaggerated in humans and no other animal is dependent upon culture like we are. Culture is usually defined as human behavior and activities that are governed by social customs and rules, and it is perpetuated because it is passed through the generations with tradition and learning. Although chimpanzees do not have language, females in some populations are able to uphold the termite fishing culture because mothers teach their offspring how to forage for the insects. Also, the tradition for chimpanzees to crack open nuts with stones is not universal for the species, paralleling the differences between human cultural traditions.

There are several claims for the earliest art on record but the dates are best for pieces of red ochre from Blombos Cave, South Africa, at about 79 Kya. These small fragments of the soft red mineral are ground down to make a flat surface and intentionally engraved with "X's"—very much like something one might doodle in the margins of this page. There are also pierced shells that would have been used as beads for personal adornment that may date as early as 100 Kya in Algeria, Israel, and South Africa, but the dates are not confirmed.

Art becomes more prolific and achieves museum-quality through time. Some is associated with Neanderthals like simple pendants or grooved and polished bones and teeth. Sites as early as 400 Kya have preserved fragments of red and black pigments which may have been used to decorate bodies or objects. Later, after 50 Kya, it is clear that these pigments were burned and used as paint. Upper Paleolithic humans, however, left behind much more than just pigments, there are flutes, carved animals, carved women, and exquisitely painted cave murals, like at the famous Lascaux and Chauvet caves in France that incorporate senses of design, texture, and color. A great number of statuettes of so-called "Venuses" are found all over Europe and into Asia in the Upper Paleolithic. Carved from stone and ivory, these portable figures are always women and are characterized by their exaggerated anatomical features. They have enormous breasts, protruding abdomens, broad hips, and marked fat deposits on the thighs, hips, and buttocks. Few have details of the face but the Brassempouy Lady, an ivory statuette from France at 25 Kya, is one exception. Some archaeologists consider the statues to be fertility symbols, possibly reflecting a fertility goddess. Perhaps the most recognizable "Venus" is the carved Willendorf woman from Austria at about 23 Kya (Figure 5.4)

The oldest rock paintings in Africa are in Namibia and date to about 27 Kya, but those are only the ones that have been dated rigorously so far. Some of Africa's rock art may date to more than 70 Kya but they are usually overshadowed in the literature by the European cave art because of the better preservation and there are just more occupied caves in Europe in the first place. The relative abundance of art and artifacts in Europe compared to Africa in the Upper Paleolithic is due to preservation biases.

Upper Paleolithic humans left behind evidence of their elaborate body adornment as well. Some of the oldest well-dated jewelry are some

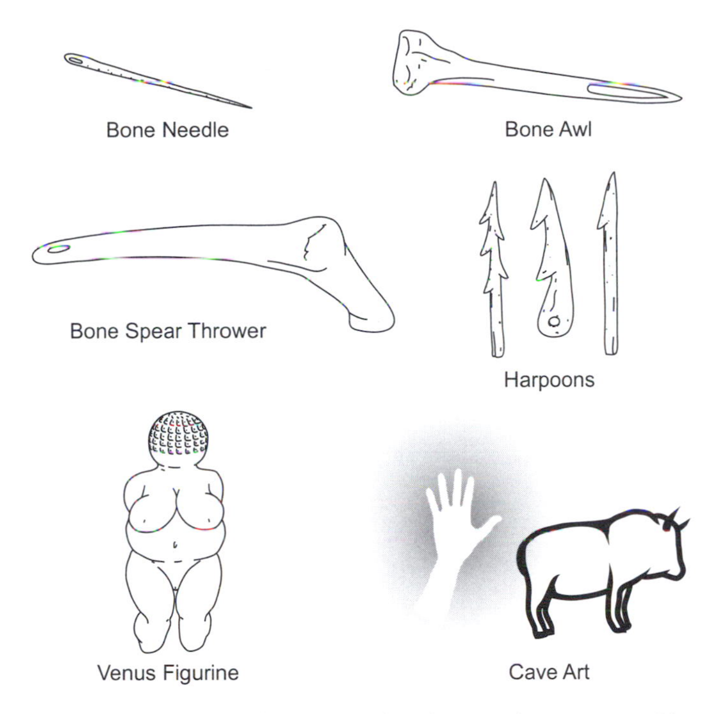

Figure 5.4 Modern human technology and art emerged in the Upper Paleolithic and includes tools made of bone, carved figurines, and cave paintings. Handprints, like the one shown, were made by blowing pigment onto the hand like a stencil. *Illustration by Jeff Dixon.*

ostrich eggshell beads with holes for stringing them together that were discovered in Turkey between 45 and 40 Kya. Around the same time, similar artifacts have been found in Kenya, Lebanon, and Bulgaria. The Châtelperronian culture is known for the use of teeth—often from carnivores like foxes, bears, wolves, and hyenas—in jewelry and body adornment.

There is no clothing preserved from the Upper Paleolithic but molecular phylogenetics of lice hold some clues. Head lice and body lice were once the same species. Head lice feeds on the scalp and body lice feeds on the body but lives in the clothing. The origin of body lice, or the timing of its split from head lice, is estimated to have occurred between 72 and 42 Kya, which means that clothing (the habitat for body lice) was probably adopted during that time. Because African body lice DNA is more variable than that of body lice around the world, it is consistent with the expansion of body lice and its clothed carriers out of Africa.

Language, as discussed above, as well as religion are traits that are traditionally lumped in with the human revolution because they involve symbolic thinking. Evidence for early religion and ritual is mostly only gleaned from burials. It is sometimes possible to tell, with controlled excavation, whether or not a skeleton has been deliberately buried or not. There is a difference in the sediment associated with the body compared to that just above and below it. Also, deliberate burial preserves skeletons much better than if they are left on the surface for scavengers, weather, and environmental processes to scatter or destroy the bones. Deliberate burial need not necessarily be equated with ritual or religion; it could merely be a way of housekeeping. Evidence of intentionality is claimed when a skeleton is found tucked in a fetal position, but critics argue that convenience (a smaller grave) rather than ritual could be responsible.

The intentionality of human burials is not debated. They occur in open air and cave sites and include beads, tools, ocher, charcoal, and other objects. It is possible—and to some it is doubtless—that Neanderthals deliberately buried their dead too. There is evidence supporting Neanderthal burial at sites like Shanidar in Iraq (60 Kya) where flower pollen was excavated from around the remains of the Neanderthal skeleton. The pollen could indicate that Neanderthals performed rituals and possessed some reverence for life and death. Critics argue that the pollen could have accumulated accidentally and is not necessarily the result of ritual or the belief in an afterlife. Also as an aside, judging from the trauma to the Shanidar skeleton, the individual had clearly led a difficult life. A heavy blow to the left side of the skull had most likely left him partially blind and his arm was partly amputated and healed. A sense of prolonged loss may have begun in humans as early as 160 Kya since it is clear from the polished skulls found at the Herto site in Ethiopia that someone had kept them and handled them long past death.

ALTRUISM AND THE HUMAN COLONY

Kindness is not necessarily a uniquely human trait, but humans do seem particularly apt to perform selfless acts at their own expense. Natural selection predicts the evolution of cooperative (benefit to all) and selfish (benefit to individual, cost to someone else) traits because both increase the fitness of the individual who carries them. Altruistic (cost to individual, benefit to someone else) and spiteful (cost to everyone) traits decrease the fitness of their carriers and one predicts that natural selection would not favor those behaviors. Thus, altruism is an evolutionary puzzle; it is a behavior that seems incompatible with natural selection.

How can selection favor costly behaviors, especially those that hinder or actually prevent the individual from successfully reproducing?

The phenomenon of altruism is documented in many animals, especially insects like worker bees. Cape hunting dogs and Florida scrub jays are just two species in which a significant number of adults do not breed and instead they help others rear more offspring than they could possibly handle on their own.

Group selection theory predicts that altruistic traits should spread. Genes for altruism would be perpetuated in future generations because altruism is good for the group as a whole. (Genes for altruism do not necessarily directly cause an individual to act selflessly, they could, for example, code for a rewarding, feel-good natural dopamine high, or stress reducer, that kicks in when an individual behaves altruistically.) The problem with this logic is that a trait will only spread if the genes that shape it are good at getting into the next generation. Altruism genes are bad at getting themselves into future generations. On top of that, altruistic individuals add to the fitness of competing individuals as they are decreasing their own. Under group selection theory, altruists would be selected against. In general, advantages for the group are generally not sufficient to spread a trait.

A combination of natural selection and kin selection solves the altruism conundrum, at least when kin are benefiting. Kin share genes so selection should favor altruism that is selectively aimed at relatives. In this scenario, shown by W.D. Hamilton in 1964, selfishness would be disfavored and cooperation would be favored. An altruist may have low direct fitness, but their overall or "inclusive" fitness would be high if they aided in the successful survival and reproduction of their gene-sharing kin.

Reciprocity, however, is the key to nonkin altruism. More specifically, reciprocal altruism, a concept introduced by Robert Trivers in 1971, can explain the behavior when two conditions are met: 1) costs must be small compared to benefits, and 2) the altruist and the recipient must interact frequently and regularly exchange roles. Such a system of tit for tat turns short-term altruism into long-term cooperation and continues as long as the benefit is always greater than the cost for each actor.

A stable reciprocal altruistic relationship is equivalent to a friendship. The actors do not need to be related for this system to work as long as it is balanced. Many primates like baboons will form male–male coalitions against higher ranking males. Often grooming another individual will be considered like a down payment for future help if attacked by a threatening individual. Chimpanzees that groom one another are more likely to share meat from colobus monkey kills with one another.

Reciprocal altruism works as long as it is withheld from cheaters who do not participate fairly. So cheater detection is crucial and the ability to discriminate between reciprocators and nonreciprocators is essential. Cheaters must be either avoided or punished. Plus, the actors must also be able to track the benefits and costs of altruist acts. This is probably where increased human cognitive functioning comes into play since humans are far better at long-term and complex networks of reciprocal altruism than chimpanzees, and far more dependent on it as well.

Large-scale societies based on agricultural resources that popped up after 10 Kya are relatively new. For most of human evolution, groups were small and subsisted on hunting, gathering, and foraging but the need for the neural wiring to manage multiple networks of reciprocal altruism across space and time was clearly important even in small groups. Once sedentary lifestyles, based on crops were adopted by some human populations, people were able to thrive in large-scale societies thanks, in part, to the altruistic behaviors that were selected for in their ancestors.

Humans group together with extended families and kin with larger groups on the order of 100–1,000,000 and this is a typically peaceful arrangement because the basis of human sociality is reciprocity (cooperation, coalitions, and exchange), which is rare in nonhuman primates even though most primates live in groups.

There are different reasons primates evolve to live in groups. Group living behavior can be driven by the need to defend resources (strength in numbers) or the need for protection from predators (safety in numbers). But groups can only evolve if the benefits to the individual outweigh the costs. (Remember, selection acts on the individual, not the group or the population.) Benefits to the individual include access to food (e.g., group hunting), access to mates, decreased predation, and communal offspring care. The costs of group living include competition for resources, competition for mates, predator risk, disease risk, and parasite risk. As a consequence, dominance hierarchies form within groups with some individuals having higher or lower rank or status than other individuals.

WAR OR PEACE?

Unlike any other primate, it is possible that humans live in groups because of the need for predator protection from members of their own species. Although other primates show violence toward their own species, humans are exceptionally good at killing one another. Certainly when driven to extremes like during time of draught or on islands where resources are quickly outstripped, humans will attack and eat their own species. Numerous archaeological sites around the world have strong evidence for human cannibalism. Even a 43 Kya Neanderthal site in Spain at the El Sidrón Cave shows that several (up to eight) skeletons, who were already victims of poor health and poor nutrition (as evident from the stress lines called "hypoplasias" in their tooth enamel), were butchered, dismembered, and eaten. Cannibalism is not uncommon under duress and is a widespread phenomenon in animals. Human warfare and genocide, however, is uncommon, but can we blame it on our ancestors?

We share an evolutionary history with primates that practice infanticide as a successful adaptation. Male gorillas and gray langurs (Old World monkeys) that are new to a group of females will kill the infants so that they can start their own lineage as soon as possible and increase their own reproductive success.

Our even closer relatives, common chimpanzees are highly territorial. Males patrol perimeters and boundaries and will attack and sometimes kill members of neighboring groups. Chimpanzee society is characterized by male hierarchical relationships which is different from that of bonobos despite their recent evolutionary split just over 2 Mya. While chimpanzees resolve conflict with politics and violence, bonobos ("pygmy" chimpanzees) make peace with love and sex. They have a more female-oriented society where females cooperate unlike chimpanzees. Bonobos will greet rival groups with genital handshakes and sensual body rubs to avoid conflict and if there is conflict, it is resolved swiftly with kissing and sex in various positions between males and females, female and females, and males and males.

Is there any way of knowing which species best approximates the LCA? Such mental exercises should be undertaken with caution since much less is known about the small populations of bonobos in the wild than about chimpanzees. With increasing field observations of bonobos, it is becoming clearer that the two species are more similar than we once thought. Prior to now, much of the peace-loving behavior of bonobos was known only from captivity where food is always abundant. Furthermore, our tendency to identify with bonobos because of their alleged tendency toward bipedal behavior is now in question since some scientists like Bill McGrew have shown that bonobos and chimpanzees perform equivalent bouts of bipedalism.

Evidence of war or peace from further back in the prehistoric record has been interpreted in various ways. Working among fossils every day it became abundantly clear to Raymond Dart how many parts of the skeleton would make good weapons. Bovid jaws with jagged teeth for cutting, large heavy bones destined for blunt force, fractured long bones that flake like stone and can be as sharp as a stone point are all parts that could and would do serious harm. The abundance of seemingly great weapons at Swartkrans Cave in South Africa led Dart, in his early years during the first half of the 20th century, to speculate that the hominins that lived there had a violent culture. If not to kill prey, they were using the bones of the prey to threaten, maim, or kill one another. This hypothesis came to be known as the "killer ape" hypothesis and their toolkit was named the "osteodontokeratic" culture. The idea lost support through the years with increasing understanding of taphonomy.

Take away the misinterpreted weapons and australopiths were not particularly threatening. Although they could have been just as strong as chimpanzees, they had much smaller teeth. Plus, there is no direct evidence they were making stone tools until 2.5 Mya (if they were the species making them in the first place). Lacking sharp teeth, size, and technologically advanced weapons, australopiths probably used brains, agility, and social skills to escape from predators. Once overarm throwing evolved and action-at-a-distance was possible, not only could hominins chase away predators and competitors for meat but cheaters (in reciprocal relationships) could be punished with little cost to the punisher (as opposed to using hand-to-hand combat). Throwing, undoubtedly, permitted warfare as well once tool technology, like spear-throwers invented by 30 Kya, enhanced the distance and force applied to projectiles.

6

Beyond Human Origins

MULTIREGIONAL AND OUT OF AFRICA MODELS

Homo erectus was the earliest hominin to disperse outside of Africa. Until then all hominin fossils from about 6 Mya to 2 Mya are restricted to Africa. Almost as soon as *H. erectus* appeared on the savannahs of East Africa around 1.8 Mya it spread as far north as Dmanisi, Georgia, and as far east as Java, Indonesia. *H. erectus* endured in parts of Indonesia until as late as 30 Kya. This massive geographic and temporal range of *H. erectus* is the inspiration for the two major rival theories of modern human geographic origins and dispersal.

Although many paleoanthropologists' hypotheses fall somewhere in between, the two models for modern human origins are known as the Out of Africa (OA) or "Replacement" or the "Garden of Eden" model and the Multiregional (MR) or "Regional Continuity" or "Trellis" model (Figure 6.1). Despite the names of the models, both agree that the roots of human evolution lie in Africa. They just disagree on the timing and the amount of evolutionary participation from *H. erectus* and Archaic populations around the world. Currently more scientists favor OA because of the convincing genetic evidence, but neither model has been conclusively supported to the exclusion of its rival, nor has either been conclusively refuted.

The strict MR model, which is supported by Milford Wolpoff and others, is based on the hypothesis that the hominins that first dispersed out of Africa evolved in different geographic regions and then interbred with modern humans as they spread across the Old World. In other words, *H. erectus* and Archaic humans (e.g., Neanderthals and *Homo heidelbergensis*) spread to Europe, Asia, Indonesia, and then, in the respective regions, evolved into *Homo sapiens*. Under MR, populations in

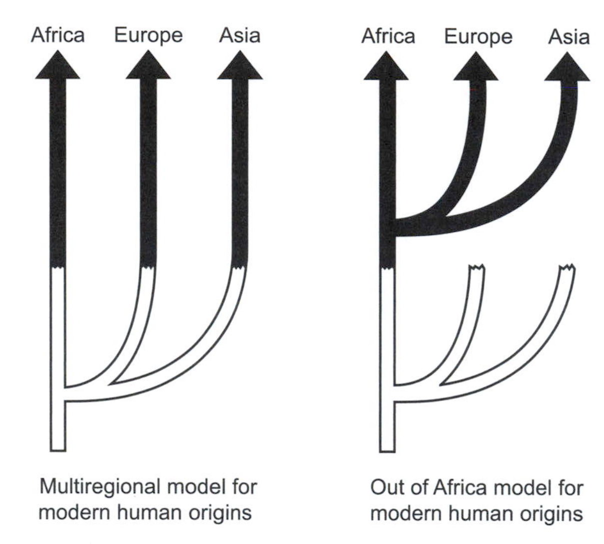

Figure 6.1 The two extreme, rivaling models for the origin of modern humans both agree that a *H. erectus* or Archaic ancestor evolved in Africa and then dispersed around the Old World. But, the Multiregional (MR) model holds that the ancestor evolved into modern humans, while the Out of Africa (OA) model posits that those early populations faded away and were replaced by a second, modern human wave from Africa after 200 Kya. *Illustration by Jeff Dixon.*

different geographic regions were linked with gene flow like they are today, but that gene flow began 1.8 Mya.

Evidence supporting this hypothesis is mostly morphological. MR points out that like modern humans, extinct hominins show geographic variation and adaptations and supporters link that regional anatomy of hominins to modern humans living in those regions today. Modern Chinese populations retain features similar to fossil *H. erectus* and Archaic humans from the region like shovel-shaped incisors and certain features of face and skull shape. Comparisons are also made between *H. erectus* and indigenous Australians. Also, modern Europeans are argued to retain similarities with Neanderthals like nose size, body size, stature, and robusticity.

Tantalizing evidence for interbreeding between Neanderthals and modern humans also supports MR. The discoverers of a child's skeleton in Lagar Velho, Portugal, suggest it is a "love child" because of its mixture of Neanderthal and modern human features. The skeleton, buried with pierced shells and red ochre from Lagar Velho, Portugal dates to about 25 Kya and has Neanderthal-like limb proportions with a human's chin. Critics are not convinced of the hybrid claim, however, and argue that the proportions lie within the human range, albeit on the cold-adapted end of the spectrum. Erik Trinkaus sees mosaic features in other European fossils as well, like from Romania around 40 Kya. Unfortunately, there is hardly a way to prove that Neanderthals and modern humans did not interbreed without finding, for example, a Neanderthal skeleton buried in the same grave as a modern human. Perhaps as the map of the Neanderthal genome grows and as understanding of gene functions increase, geneticists will be able to determine whether or not there would have been molecular barriers to their interbreeding with humans.

The strict OA model holds that after hominins dispersed from Africa first as *H. erectus* and Archaics, a second major dispersal of modern humans replaced those existing populations around the world with no genetic mixing. (Although looser interpretations of OA allow a small amount of gene flow.) Replacement could have been the result of indirect competition, like simply out-surviving the more primitive groups, or direct competition like warfare and genocide. Strict OA includes no interbreeding between modern humans and Archaics and argues that modern human geographic diversity occurred recently after 200 Kya.

Fossil and archaeological evidence support OA since the oldest modern humans are found in Africa (Omo I and II). The anatomical similarities across modern human populations are exceedingly closer than the similarities between premodern and modern humans within a particular geographic region. Furthermore, if Archaics were evolving into modern humans, there should be no overlap in their existence, yet at the Israeli sites of Skhul and Qafzeh from 110 to 90 Kya, modern humans were living very close to Neanderthals at the nearby site of Tabun around 110 Kya. The first modern humans arrived in Europe by 40 Kya and had tropical adaptations in their body proportions, but Neanderthals were still evolving there until 28 Kya with their adaptations to the cold.

After 40 Kya there are many modern human occupational sites across Western Europe. The oldest evidence comes from sites like Bacho Kiro and Temnata, Bulgaria, (43–40 Kya), Peştera cu Oase, Romania, or "cave with bones"(40–35 Kya), then Kent's Cavern in England (30 Kya). The overlap between modern humans and Neanderthals in Europe lasted about 10,000 years before Neanderthals disappeared. Some proponents of OA have blamed humans for causing the Neanderthals' demise. As discoveries accumulated in the Middle East and Europe a clear pattern emerged: modern human and Neanderthal sites remained distinct, but Neanderthal sites became fewer and farther apart through time.

Most of the morphological evidence so far for the MR model is not nearly as persuasive as the molecular evidence that tends to support the OA model. The dispersal of modern humans out of sub-Saharan Africa occurred after 200 Kya according to both mitochondrial DNA and Y-chromosome analyses. Ancient DNA (aDNA) evidence also supports OA. Neanderthal DNA lies outside the normal range of variation found in modern humans, but it is not as divergent as chimpanzee sequences are from modern humans. The genetic distance between Neanderthals and modern humans, which is the same even for modern Europeans, suggests Neanderthals did not evolve into modern humans or contribute in an ancestor–descendent fashion to modern human origins. Even aDNA from fossils that have what some researchers call transitional anatomy is not transitional.

Race

H. sapiens is one variable species that is held together through large-scale gene flow. Between populations with contrasting allele frequencies there exist populations with intermediate frequencies. This phenomenon is known as a cline, or an allele frequency gradient in space held together by gene flow. Human variation represented by a cline is continuous in spectrum and there are no discrete groups. Therefore, the concept of "race" as is used today has been deemed horribly flawed and should be replaced with a notion of continuous geographic diversity in biology and culture.

Although, the modern scientific mantra that "there is no biological basis for race" is true on many levels, it is also misleading. Obviously, there are regional phenotypic differences among populations of people and these are evident in, for instance, skin color, hair color and texture, eye color and shape, nose shape, height, limb lengths and body proportions, skull and face shape, ear wax type, blood type, disease susceptibility and resistance, body fat distribution, high altitude adaptations, lactose tolerance, and so on. These differences are caused by random genetic drift, sexual selection, differing levels of gene flow, and of course, regional environmental adaptations. Regional variation also records ancestral history. The eye shape and hair texture of Native American Indians reflects their ancestral Asian origins. The traits we use to identify "races" are nearly always continuous traits that exist to varying degrees in all human populations.

Nuclear DNA studies of geographic variation have shown that the traits that reflect regional variation are determined by a tiny fraction of our genes.

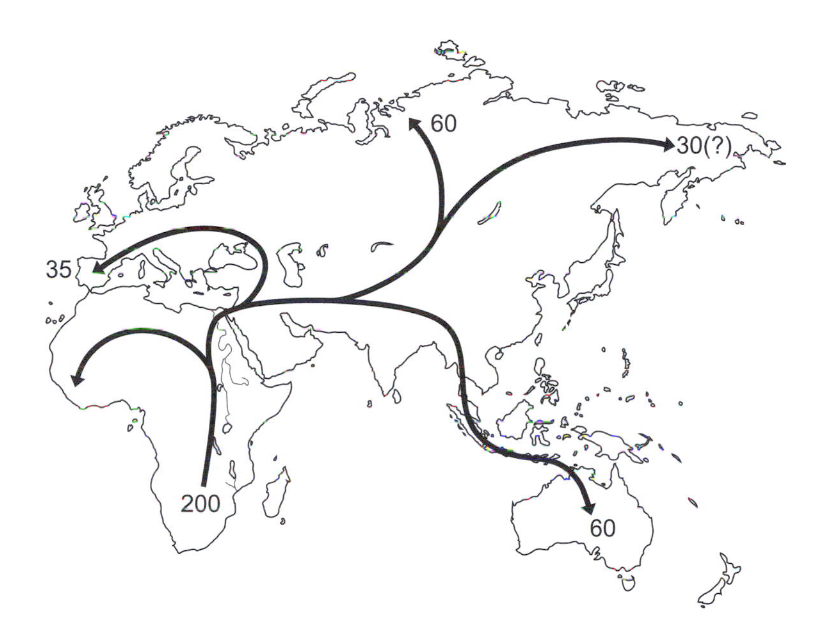

Figure 6.2 Arrows on the map of the Old World show approximate dates for the first arrival of modern humans to geographic regions, including the first dispersal to the New World (the Americas) by about 30 Kya. *Illustration by Jeff Dixon.*

Plus, an underwhelming 15 percent of human genetic variation separates major human groups. Compared to various species of large-bodied mammals with excellent dispersal abilities (e.g., the gray wolf and white-tailed deer), humans have very little variation. Most human diversity exists as differences among individuals within populations. That is, there is more variation within populations than between them.

WORLDWIDE DISPERSAL

The peopling of the planet should be considered a consequence of dispersal rather than intended migration. It should be imagined as a fluid process with people moving back and forth as opposed to a singular, directional colonization event (Figure 6.2).

The cause of the initial hominin dispersal out of Africa is not explicitly told in the fossil or archaeological records. An early hypothesis suggested that the invention of the Acheulean hand ax enabled *H. erectus* to inhabit new territories. It could, in a sense, afford to venture out with its new technology. The idea is based on the association of the ubiquitous hand ax with *H. erectus* and the presence of hand axes outside of Africa, The assumptions behind this idea are too transparent for most scientists today considering elephants dispersed from Africa without the benefit of hand axes.

Something more biological as opposed to cultural was probably the cause of *H. erectus* dispersal. A predatory species cannot survive in large groups. Since it sits atop the trophic pyramid, a predator cannot outnumber its prey or it will starve to death. Pat Shipman and Alan Walker proposed that in order for *H. erectus* to make the carnivore transition, it probably either reduced its population size or increased its geographic range and then also increased its body size. The fossil record indicates the latter two clearly occurred after 1.8 Mya, just about when evidence for meat-eating becomes prevalent. The Old World dispersal of *H. erectus* was probably a consequence of becoming a predator and simply following the herds.

The cause behind the dispersal of modern humans could have been the result of similar pressures. Predatory issues were most certainly an issue for early humans, however by the emergence of *H. sapiens* we have the first clear-cut evidence of fire, shelter, a higher degree of sociality, a brain size increase, enhancement of technological skill. The evolution of technology and adaptability probably facilitated the expansion of humans into more marginal and previously inaccessible environments compared to *H. erectus*. There is also abundant evidence that modern humans in various regions were incorporating a variety of prey items like tortoises, hares, ground birds, and shellfish into their diets, so not all populations were necessarily following herds.

Sites along the world's coastlines suggest that early human populations were successful at living on shores and this lifestyle probably facilitated dispersal especially by the time boats were ubiquitously used. The origins of boats and seafaring crafts are unclear at present but it is generally safe to assume that the first Australians between 60 and 50 Kya had such inventions. The colonization of Australia is intriguing because even at low sea levels there was never a land connection between the Sunda continental shelf of Northwest Indonesia and the Sahul continental shelf, which contains the Australian landmass as well as Papua New Guinea and Tasmania. Australia is an unusual case because the continent may not have been visible from any Indonesian islands, even at times when sea levels were lowest during glacial periods when water is tied up at the frozen poles. Although bird flight patterns could have indicated land beyond the horizon, the distance is considered too far for people to simply island hop the way *H. erectus* (and maybe *Homo floresiensis*) did throughout Indonesia. To get to Australia, it is assumed that people would have had to build boats or crude rafts and endeavored out onto the open sea to explore. Getting to Australia probably required technical seafaring skills as well as the ability to plan, work as a group, and probably talk to one another. Once on the big island, there is evidence that they moved to the southern part rapidly. By 23 Kya at Willandra Lakes in southeastern Australia, early inhabitants left behind the largest collection of Pleistocene human footprints in the world. Fossil skulls and skeletons from the earliest sites like Kow Swamp in Northern Victoria and Lake Mungo show that only modern humans were able to reach Australia.

The first Americans were also probably seafaring people. They came from northeast Asia and were not just good at the marine lifestyle but that they also adapted to an arctic one. Genetic analyses estimate that humans first arrived in the Americas as early as 30 Kya, but more likely between 25 and 15 Kya during the height of the last glacial period. The most probable route was across the Bering Strait from Siberia to Alaska which, when sea levels were lower, would have been largely closed and some Aleutian islands would have been linked, making the journey (although not necessarily an intentional one) more terrestrially anchored and therefore less treacherous.

Jon Erlandson suggests that the first Americans could have grazed their way from Asia to the America on an "ancient kelp highway." According to the "coastal migration theory," they may have lived in small maritime populations that boated from island to island, or shore to shore, hunting the sea creatures that lived in kelp forests. At present there is a nearly continuous kelp highway from Japan and Siberia, across Bering Strait to Alaska, and down the California coastline. And in kelp forests live, for instance, seals, sea otters, fish, and sea urchins. Some of the earliest archaeological sites in the Americas are found near productive kelp forests. Daisy Cave in the Channel Islands off southern California, dated to nearly 10 Kya, has preserved evidence that humans used kelp resources there.

The Clovis culture, named for its discovery in Clovis, New Mexico, is known for its elegantly distinctive spear points and dates to 11 Kya at the end of the last Ice Age. The Clovis people were long given the title of the earliest humans to arrive in America, but the geography of most sites is skewed to the eastern side of North America, while earlier and contemporaneous pre-Clovis sites of human occupation have been found in the west and in South America. Duktai in Alaska is about

12 Kya, and others like the Meadowcroft rock shelter in Pennsylvania, Topper in South Carolina, Taima-Taima in Venezuela, Pedra Furada in Brazil, and Monte Verde in Chile are as old and could be even older. Meadowcroft in particular may be as old as 20 Kya. By 12.5 Kya in Monte Verde, Chile, archaeologists found the earliest permanent settlement in the New World with the remains of a shelter that would have been large enough to house upwards of thirty people. Just like the dispersal out of Africa, there were likely multiple influxes of human populations into the New World and the continued discovery of archaeological sites as well as the deeper analysis of genetics of living human populations will help recount the true history of the first Americans.

Did Neanderthals Discover America?

While screening the dirt and rocks from a creek bed looking for evidence of the Santee Canal built in the 1790s (30 miles west of Charleston, South Carolina), underwater archaeologist Mark Newell recovered a crude stone tool. It was a remarkable find considering the tool is a type that was made by Neanderthals in Europe. Plus, the raw material was obviously foreign. Microscopic invertebrate fossils in the rock confirmed that it originated in France. In fact, the unique geochemical composition of the stone pointed to its source in a particular rock formation in the Bordeaux region. Could Neanderthals have floated across the Atlantic in the Stone Age to discover North America?

It would be tempting to conclude Neanderthals were the first Americans, but with further evaluation a simple explanation is clear. In relatively recent history, European rivers that hold prehistoric artifacts were mined for ballast, and then European ships dumped their ballast in American ports when taking on cargo for the voyage back home. These dumps were then mined for ballast that was recycled in local American watercrafts, like those that traveled in the Santee canal.

WILL WE EVOLVE OR WILL WE GO EXTINCT?

There are tales born from misunderstanding evolution that humans will eventually lose their diminutive fifth toe or their rupturing appendices. Men are often projected to lose their milkless nipples and women are forecasted to develop extra-large or even additional breasts to accommodate exaggerated sexual preferences. It is even commonly assumed that the world's population will eventually share a single skin tone, even though we know from both life experience and genetics that skin color inheritance is not as simple as mixing each parent's pigments together.

These speculations lead naturally to the question: Is natural selection working on us anymore? From examples like lactose tolerance, we know that evolution has brought about major changes to human populations within just the last several thousand years. Resistance to HIV infection and malaria is higher in some populations than others. Clearly human cultural adaptations and innovations are not preventing us from developing some biological ones. But if natural selection is currently acting in human populations, does that mean speciation is on the horizon? And can we dare to ask such a thing without flirting with racism?

With modern modes of global travel and with decreasing cultural barriers between populations (languages are going extinct all over the world), gene flow is most likely going to prevent any new species from branching off of humans. With 6 billion of us on the planet, it seems unlikely that any human population could become so reproductively isolated that it splits off into a separate species, but such a scenario is not outside the limits of reality especially in the wake of a hypothetical worldwide catastrophe like the dinosaurs experienced. Fortunately NASA is constructing plans for diverting an asteroid like the one that caused the dinosaur extinction should one threaten Earth again.

Avoidable or unavoidable catastrophes aside, can we use medicine to keep us from going extinct? Medicine along with agricultural technology will probably help keep *H. sapiens* from a predictable demise, but they cannot help us evolve into more extinction-resistant forms. Unless medical changes are made to the gametes and inherited in future generations, all the best organ regrowth and antiaging treatments will mean next to nothing in evolutionary terms. But natural selection could still work on fertility or fetal survival and sexual selection can shift things as well. There is evidence that a number of our genes involved in lactose tolerance, brain development, skin pigmentation, reproductive organ development, metabolism, and disease resistance are undergoing strong selection right now which will most likely lead to changes at least at the population level in the future. As a species, we will continue to evolve into the future, but how much?

We could evolve into forms so different from our current ones that future biologists and paleontologists would call us a different species. In that sense, *H. sapiens* as we know it now would be extinct. We may not have much longer to evolve, however, because the next big faunal turnover, or cyclical extinction event, is predicted to occur at around AD 500,000. On average, mammal species last about 2.5 million years, which is the duration between turnover events. It is probably no coincidence that drastic evolution in the hominin lineage corresponds to the last turnover event around 2 Mya. These extinctions are spaced by intervals of climate change (affecting temperatures, precipitation, habitats, and food sources) that are controlled by astronomical variations like changes in the Earth's orbit and its wobbling on its axis (Milankovitch cycles). 500,000 years is plenty of time to evolve into something new, after all, 500 Kya there were only moderately intelligent Archaics and *H. erectus* on Earth and look where hominins stand today.

But while we thrive in a population of 6 billion, our closest relatives are on the brink of extinction. The World Wildlife Fund (www.worldwildlife.org) reports that great apes are all endangered. There are only about 10,000 bonobos left in the wild, down from over 100,000 just twenty years ago. They are being squeezed out of preferred habitats by humans and are also being killed for bush meat to be served at expensive restaurants. Orangutans are stuck on islands that are becoming increasingly deforested. Their numbers are dwindling at 50,000. Jane Goodall reported on her Web site (www.janegoodall.org) that there were only 150,000 chimpanzees left in 2004, but that there should be at least be a million today. Less than 1,000 mountain gorillas have survived to the present day with no help from poachers and deadly Ebola outbreaks.

Witnessing the inevitable wild extinction of our closest relatives reminds us just how adaptable we are, but just because we normally tell one single evolutionary tale (our own) does not mean that all of evolution culminated in the result of humans. Although, for many people, being human means being elevated from the fishes and the rest of the animal kingdom, simply deciding that humans are the king of the planet does not mean we actually are. Granted humans cover nearly the entire dry surface of the world and some of the wet surface as well, but there is only one species of upright apes alive today after a mere 6 million years of evolution, yet there are over 350,000 species (and counting) of beetles which have been evolving for over 260 million years. In that regard, beetles are the king of the planet having successfully radiated and propagated far more than humans ever have and ever will.

DEEP IMPACT

Perhaps the uniquely human trait with the most impact on the world is our propensity for changing our environment. We are experts at creating trash and we leave it behind everywhere in heaps and mounds. We also cannot travel far without our flocks of cattle, sheep, and goats that are numbering over 3 billion on the planet. We plough up fields, tear down forests, and light up the night sky. Wherever we go the native flora and fauna change. Some capitalize on our arrival, like the rats and birds that thrive on our garbage, but others shrink away and sometimes disappear for good.

Humans are so capable at manipulating the environment that we have even changed the global climate. Certainly global warming is a natural process caused by glacial cycles, but burning fossil fuels emits pollutants like carbon dioxide into the atmosphere where it gets trapped and causes "global warming." Scientists expect that within the next 100 years, temperatures may rise to 10 degrees Fahrenheit and sea levels could rise upwards of 20 inches. The world will look much different than it does today in just a century's time.

The polar ice caps are melting and threatening the livelihood of big mammals like polar bears whose hunting strategy depends on floating ice for catching prey like seals in the water beneath. Very slight changes in temperature affect smaller animals that are incapable of adapting or moving around with the climate. For example, tree frog species are dying off because those that live on mountain tops move steadily toward the peak until they can no longer find cooler temperatures.

Humans have a history of causing animal extinctions or at least being held responsible for them. For example, the disappearances of ground sloths in the West Indies and pygmy mammoths on Wrangel Island in the Arctic Ocean are blamed on human hunters. Extinctions of small and especially large animals (megafauna) occurred in the Americas between 12 and 10 Kya, in Madagascar between 6 and 1 Kya, and in Australia between 40 and 30 Kya and it is still debated whether they were induced by the arrival of humans to the regions or if they were spurred by climate change. In North America, the megafauna (including mammoths, mastodons, giant beavers, wild horses, camels, and saber-toothed cats) vanished from the fossil record just about the same time some of the earliest artifacts like Clovis points appear. Critics argue that the best evidence to show that overhunting by humans caused the extinctions would be an abundance of kill sites (butchered skeletons with spear points), but there are not an inordinate number of these. However, it is possible that climate change from regular glacial cycles combined with the threat of a cunning human predator could have been a deadly duo. Perhaps by overkilling grazing animals, humans indirectly caused environmental changes in the landscape, which affected the other animals. Today, ecologically friendly hunting patterns are managed by local, state, and federal government institutions in the United States.

Hunting and fishing regulations may not be enough to stop us from repeating our past mistakes, however. Because of overfishing and pollution, scientists have predicted that unless drastic changes are made, within the next fifty years nearly all marine species of seafood are going to collapse. Ironically, humans may be responsible for killing off one of the crucial food sources that contributed to making us "human," since incorporating marine animals into our Paleolithic diet could have influenced the evolution of our brain.

REWINDING AND REPLAYING EVOLUTION

What are we left to think about human existence? With all the evidence pointing to our ape-like origins 6 Mya and our shared ancestry with everything down to cockroaches, slime molds, pond scum, tarantulas, pit vipers, foot fungus, and even viruses, the unique evolutionary status of *H. sapiens* may seem bleak and maybe even disgusting.

But even though we have lousy relatives, should we think less of our species? If we can forget about the aforementioned negative impacts we have made on the planet is it possible to still revel in our exquisite uniqueness? After all, it is remarkable what Mother Nature made from an ancestor with a small brain, clumsy hands, and no significant emotional or artistic forms of expression. We are both odd and wonderful, but are we unique for all time and in all of eternity? What if just one event happened differently in the course of Earth's history—would humans still have evolved?

If an asteroid had never wiped out the dinosaurs it is possible we would not be here today. The worldwide extinction event at the Cretaceous-Tertiary boundary spared small mammals and it is from these survivors that the entire mammalian radiation was born. Thanks to those early mammals, it is giraffes, not brontosaurs, that eat the leaves from the tallest trees and it is humans, lions, and hyenas, not tyrannosaurs, that sit atop the trophic pyramid.

Consider setting the clock back to zero, back to the Big Bang and the creation of this universe 14 billion years ago. Then set it in motion. What would the "redo" of history look like? Would it be the same, like a television rerun, or would it become a completely different episode?

Paleontologists Simon Conway Morris and Stephen Jay Gould (1941–2002) entered a famous debate on this issue. Simon Conway Morris argues that a redo of evolution would be a rerun because history is constrained by laws of physics and biology, and so forth; not all things are possible. Under normal environmental forces, life will adapt accordingly and there are only so many ways to do that. Morris emphasizes the numerous examples of convergent evolution like sharks and dolphins, and the evolution of saber-toothed cats in both the marsupial and placental mammal lineages.

Stephen Jay Gould saw the complete opposite view and argued that a replaying of evolution would result in a whole new episode of life. He called humans and the entire guild of living organisms alive today a "glorious accident" brought about by a unique unrepeatable sequence of events during Earth's history. Each time evolution is rewound and replayed and the lottery of events starts all over, the result will be an entirely new outcome with entirely unique organisms. Although he concedes that there are limits to the diversity that could result from such a hypothetical experiment, Gould argued that the chances are nearly impossible that today's world with today's creatures would inevitably evolve if Mother Nature was given another whirl. To Gould, examples of convergence are rare and underwhelming and the most important feature at stake in his argument, the uniqueness of humanness, humanity, and human consciousness, has not evolved in any other living being. From Gould's perspective, it is hard to imagine that anything like *H. sapiens* would evolve ever again.

APPENDIX A

HUMAN AND CHIMPANZEE SKELETAL ANATOMY

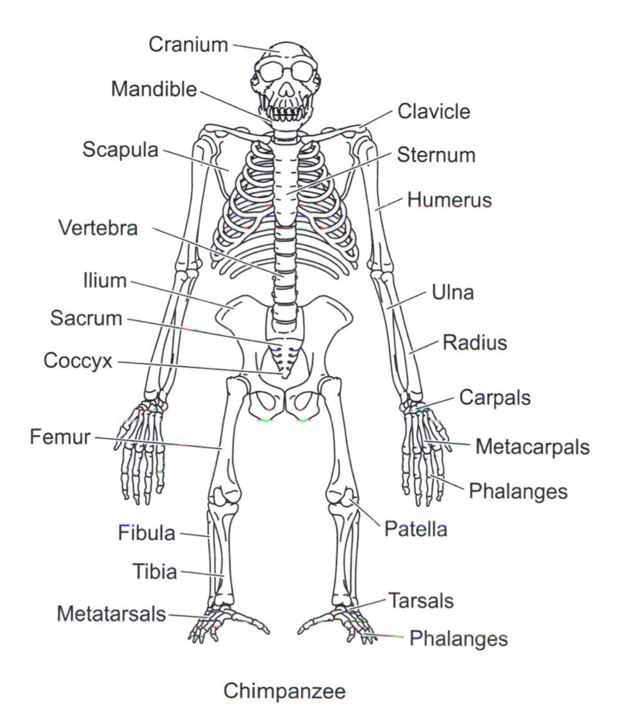

Chimpanzee

Illustration by Jeff Dixon.

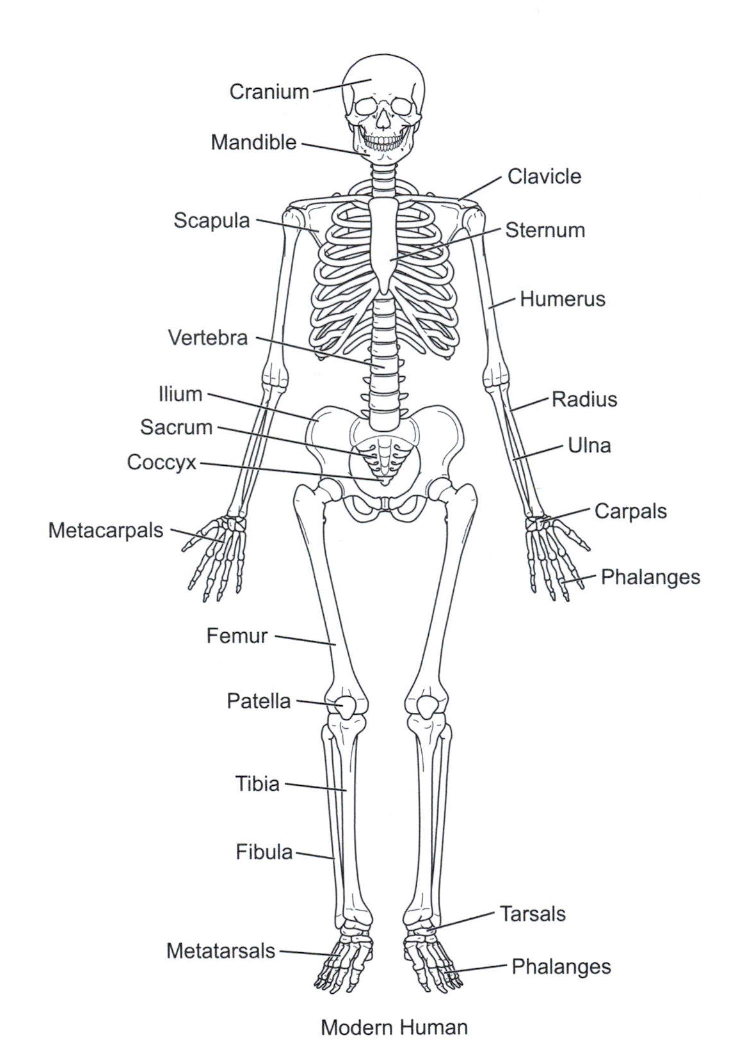

Modern Human

Illustration by Jeff Dixon.

Appendix B

Geologic Time Scale

Era	Period	Epoch	Millions of Years Ago (approx.)	Event Highlights
Cenozoic	Quaternary	Holocene	0.01	The end of the last Ice Age and the beginnings of agriculture, animal domestication, and large-scale civilization
		Pleistocene	1.8	Ice Ages Dispersal of hominins out of Africa
	Neogene	Pliocene	5	Emergence of bipedal hominins Monkeys radiate Large carnivores make their presence
		Miocene	23	Apes evolve and radiate
	Paleogene	Oligocene	35	Monkeys and large running mammals emerge
		Eocene	54	Radiation of modern mammal types
		Paleocene	65	First placental mammals First primates
Mesozoic	Cretaceous		145	First flowering plants Dinosaurs climax and then go extinct at the end of the Cretaceous
	Jurassic		200	Dinosaurs are abundant First birds. First mammals

(continued)

(continued)

Era	Period	Epoch	Millions of Years Ago (approx.)	Event Highlights
	Triassic		250	First dinosaurs. Conifers abundant
Paleozoic	Permian		290	At the end of the Permian trilobites go extinct in an enormous extinction event
	Carboniferous		350	First reptiles. Impressive coal forests
	Devonian		400	Sharks and amphibians dominate and are abundant
	Silurian		440	First terrestrial plants and animals
	Ordovician		500	Invertebrates are dominant. First fishes
	Cambrian		540	"Explosion" of animal life in the seas, includes trilobites
Precambrian			4,600	Life on Earth begins. Rare fossils of primitive aquatic plants. Some fossil algae date back to 2.5 billion years or more

Appendix C

Recommended Resources

Premier Journals for Scientific Research on Human Origins

- *Nature*
- *Science*
- *Journal of Human Evolution*
- *American Journal of Physical Anthropology*
- *PaleoAnthropology* (on-line)
- *Evolutionary Anthropology*
- *Current Anthropology*

News Media with Reporting on Human Origins Research

- *National Geographic*
- *New Scientist*
- *Scientific American*
- *American Scientist*
- *The New York Times*

Web Sites

- American Association of Physical Anthropologists: www.physanth.org
- Paleoanthropology Society: www.paleoanthro.org
- Association of American Anthropologists: www.aaanet.org
- The Leakey Foundation: www.leakeyfoundation.org
- Talk Origins: www.talkorigins.org
- Anthropology in the News: anthropology.tamu.edu/news.htm
- Becoming Human: www.becominghuman.org

- Smithsonian Human Origins Program: www.mnh.si.edu/anthro/humanorigins
- Genographic Project: www3.nationalgeographic.com/genographic/index.html
- Understanding Race: www.understandingrace.org
- Animal Diversity Web: animaldiversity.ummz.umich.edu
- The Chimpanzee "Ai": http://www.pri.kyoto-u.ac.jp/ai/index-E.htm
- The Jane Goodall Institute: http://janegoodall.org
- World Wildlife Fund: http://www.worldwildlife.org
- Genome Programs: http://genomics.energy.gov

Glossary

Acheulean. A type of stone tool industry characterized by the hand ax from the Pleistocene period in human evolution. Associated with *Homo erectus* and Archaic *Homo sapiens*. Named after the site Saint Acheul in France where it was first discovered.

Allele. A version, a sort, or a kind of a gene. For example, the gene for human earwax has two alleles, one for dry and one for wet.

Altruism. Behaving selflessly, at a cost to oneself but with a benefit to another.

Analogy. Similarity among organisms, which is the result of selection for their use in a similar function, not due to recent shared ancestry. It is the result of convergent evolution. For example, wings of butterflies, bats, and birds.

Ancient DNA (aDNA). Genetic material extracted from ancient remains that is almost always fragmented or damaged due to decomposition.

Anthropoid. Nickname for a member of the suborder anthropoidea, the so-called "higher primates," which includes monkeys, apes, and humans (but not tarsiers).

Anthropology. A broad, holistic, comparative science that is simply defined as the study of humans.

Arboreal. Tree-dwelling.

Archaeology. The study of people in the past and the cultural remains that they left behind.

Articulated. Connected. For example, when the bones of a skeleton are articulated, the thighbone articulates with the hipbone.

Atlatl. A spear thrower. The word was coined by the Aztecs but is used for all spear throwers no matter their origin, even those from archaeological sites that predate accounts of the Aztec weapons used against the Spanish.

Autosomes. All the nonsex chromosomes.

Bipedalism. Walking upright on the two hind legs. One who uses bipedalism for locomotion is said to have the trait of bipedality.

Blade. An Upper Paleolithic stone tool, which is at least twice as long as it is wide.

Bovid. A member of the family Bovidae, which is a diverse order of cloven-hoofed mammals including cows, antelope, buffalo, gazelles, sheep, and goats.

Brachiation. Locomotion by arm-swinging from tree branches. The Latin noun "brachium" means both arm and branch.

Canine-premolar (CP3) honing-complex. A system by which the upper canine is sharpened by the lower premolar when the teeth are in occlusion (i.e., the jaws are together).

Carnivore. An animal that eats mostly meat.

Catarrhine. A member of the infraorder Catarrhini, which includes Old World monkeys, apes, and humans.

cc. Cubic centimeters. *See* Cranial capacity

Chromosome. An element in a cell's nucleus that comprises DNA.

Colobine. The leaf-eating subfamily of monkeys within the Old World monkeys or cercopithecoids.

Complex trait. One that is expressed by numerous genes.

Continuous trait. One that can exist over a range of variation, like height, for instance (for opposite see Discrete trait).

Core. A rock from which flakes are struck and used to make tools.

Cranial capacity. The amount of space or volume in the skull that holds the brain and serves as a substitute for brain size when there is no brain present to measure its size directly. Cranial volume is often measured with grains of rice and is reported in cubic centimeters (cc).

Cranium (pl. Crania). The skull minus the mandible or jaw, which is often the state of fossil hominin specimens. When the cranium is missing the face and the base it is called a calotte or "skullcap."

Cusp. In dental terms, cusps are peaks on premolars and molars.

Dentition. Teeth.

Derived trait. One that is different from the ancestral or primitive form.

Discrete trait. One that can only exist in a finite number of states, for example, wet versus dry earwax (for opposite see Continuous trait).

DNA. Deoxyribonucleic acid. The molecule that makes up the genetic material in a cell's nucleus or mitochondria.

Drift. Also called "gene drift" or "genetic drift," it is a force of evolution where alleles are randomly spread throughout, or eliminated from a population in the absence of gene flow with other populations.

Ecology. The science or study of the relationships between organisms and their environments.

Endocast. Brains do not fossilize but the inside of a skull can fill with sediment. After many years, the resulting rock is a mold of the inside of the skull, resembling what the brain may have looked like during life. Scientists can also make endocasts by molding and casting the inside of skulls. Here endocast refers only to brains and skulls, but it is actually a more general term used for a variety of anatomical regions.

Evolution. Change in allele frequency through time.

Exaptation. A trait that natural selection favors for use in a different adaptation than that for which it was originally intended.

Fauna. Animals.

Fitness. Reproductive success.

Flora. Plants.

Foramen Magnum. Literally the "big hole" at the bottom of the skull for the exit of the spinal cord.

Founder effect. The result of a small subgroup of a population becoming isolated and starting a new population from a small gene pool.

Frontal. Refers to either the skull bone at the forehead or to the right and left frontal lobes of the brain just underneath it.

Gametes. Sex cells (e.g., eggs and sperms).

Gene. Unit of DNA that codes for the development of a process or trait.

Gene flow. Mating between people of different populations, or the mixture of gene pools.

Gene pool. Breeding population.

Genome. The entire genetic sequence of an individual or species (based on a composite of individuals).

Genotype. The genetic makeup of an individual. It is unique to every person.

Genus (pl. Genera). Taxonomic grouping at one level above species that usually includes animals with a similar adaptive plateau.

Great ape. The hominoids excluding the lesser apes (gibbons and siamangs), which include orangutans, gorillas, chimpanzees, and bonobos. Some include humans under the umbrella term, some do not.

Gregarious. Social or group-living (e.g., chimpanzees) as opposed to being solitary (e.g., orangutans).

Hand ax. The stone tool that typifies the Acheulean industry of the Pleistocene. It is teardrop shaped and is chipped away on both sides, also called a "biface."

Haplorhine. A member of the primate suborder Haplorhini ("dry noses") that includes tarsiers, New World monkeys, Old World monkeys, apes, and humans. The other suborder is Strepsirhini.

Herbivore. A plant eater.

Heterozygote/Heterozygosity. When the two alleles at a genetic locus are different.

Hominid. "Hominin" is widely replacing this taxonomic term for humans and their unique fossil ancestors since the split from the chimpanzee lineage about 6 Mya.

Hominin. Any living human and all fossils on the unique lineage that led to humans after the split from the chimpanzee lineage about 6 Mya.

Hominoid. Apes and humans.

Homology. Similar in structure but not necessarily in function.

Homozygote/Homozygosity. When the two alleles at a genetic locus are the same.

Hypothesis. A prediction or a provisional explanation for a phenomenon that can be tested.

Isotope. A form of a chemical element in which the atoms have one or more extra neutrons.

Kin selection. Since kin share genes, increasing the fitness of kin (e.g., by protection) indirectly increases one's own fitness.

Knuckle-walking. Mode of locomotion used by the African great apes and, to a degree, orangutans. Instead of putting the weight on the palm of the hand, like a crawling baby, knuckle-walkers put the weight on the middle phalanges of the hand. The same is not true for the foot, which hits the ground with the sole, however some apes do curl their toes under.

Kya. Thousands of years ago.

Last Common Ancestor (LCA). Also known as MRCA or the Most Recent Common Ancestor, the LCA is the shared ancestor between humans and chimpanzees before their evolutionary lineages split apart about 6 Mya.

Lesser ape. Gibbons and siamangs of Indonesia.

Life history. The developmental changes an organism experiences from conception to death.

Locomotion. The way an animal gets around and moves about.

Locus (pl. Loci). The location of a gene on a chromosome.

Long bones. The longest bones of the body, or the bones of the limbs, including the femur, tibia, fibula, humerus, radius, and ulna. Also, sometimes includes the metatarsals and metacarpals.

Macroevolution. Large-scale evolutionary changes, typified by speciation, above the population level. There is a false dichotomy between micro- and macroevolution since they are both located along the continuum of change that is driven by the same evolutionary forces.

Mandible. Lower jaw.

Maxilla. Upper jaw.

Meiosis. Division of gametes for reproduction.

Microevolution. Evolutionary change at the gene and trait level within a population. There is a false dichotomy between micro- and macroevolution since they are both located along the continuum of change that is driven by the same evolutionary forces.

Microwear. Tiny pits and scratches left on teeth by food. It can also refer to marks left on bones from teeth or stone tools, and also to marks left on stone tools from use.

Mitochondrial DNA (mtDNA). The DNA from the many mitochondria, or the "powerhouses," of the cell. It is the only DNA found outside of the nucleus.

Mitosis. Cell division everywhere in the body, except by the gametes, which divide by meiosis.

Molecular clock. A technique based on the clock-like rate of accumulation of mutations in DNA sequences, which is used to determine the time since two lineages diverged in evolution.

Monogamy. One male and one female form a pair-bond for reproducing and parenting.

Monogenic trait. One that is expressed by one gene (e.g., Mendelian inheritance).

Morphology. Technically, it is the study of the shape of something. It is also used as a synonym for shape and size. For example, tooth morphology of leaf-eating monkeys is much different than that of fruit-eating apes.

Mutation. A change in the genetic sequence in the daughter cell compared to the parent cell.

Mya. Millions of years ago.

Natural selection. A mechanism for evolution whereby favorable traits are spread to successive generations in a population by the survival and successful reproduction of individuals who have those favorable traits.

Niche. An ecological strategy; a way of making a living.

Offspring. The next generation of individuals produced by a male and a female.

Oldowan. The most primitive stone tool technology recognized by archaeologists, named for the site where thousands of early stone tools have been discovered: Olduvai Gorge, Tanzania.

Olduvai Gorge. A region in Tanzania with large-scale geologic outcrops containing fossil- and artifact-rich Pleistocene sediments. One of the "cradles of humankind."

Omnivore. An animal that eats a combination of plants, animals, and their products.

Ontogeny. Growth and development.

Opposable thumb. A thumb (digit 1) that is able to swing around, contact, and press firmly against, fingertip-to-fingertip with digits 2–5. Other primates, especially apes, can do this to a certain degree, but humans (partly because of joint mobility and partly because of finger lengths) have the most dexterous thumbs.

Paleoanthropology. The field of anthropology that studies human origins and evolution.

Paleolithic. Refers to the prehistoric times when stone tool (= "lithic") technology dominated hominin culture.

Parsimony. Follows the rule that the simplest explanation is probably the correct one.

Phenotype. An individual's physical makeup.

Phylogeny. The evolutionary history of a group or the diagram (also called a "phylogenetic tree" or just a "tree") used to illustrate that evolutionary history.

Plate tectonics. Causes continental drift.

Platyrrhine. A member of the Platyrrhini ("flat, broad noses"), which are the New World monkeys.

Polygamy. A mating pattern by which an individual mates with more than one partner.

Polygenic trait. One that is expressed by multiple genes (see Complex trait).

Postcranium. The bones of the skeleton excluding the skull. In humans this is more like "beneath-cranium" but in most animals, the skeleton is located behind the skull, hence the use of "post."

Primitive trait. One that is shared because of shared ancestry. For example, five fingers and toes are primitive for mammals since humans, lemurs, raccoons, dogs, and cats, all have them even though these groups split deep in the mammal tree.

Prognathism. The state of having the face protrude out, away from the skull. "Gnath-" means jaw and the jaws make up much of the face. Dogs have extremely prognathic faces. Adult chimpanzees are much less prognathic, but are much more so than baby chimpanzees and humans.

Quadrupedalism. Locomotor category where the body weight is carried by all four limbs.

Reciprocity. "Tit for tat" behavior. Mutual exchange or cooperation.

Robusticity/Robust. Strength, sturdiness, and thickness. Used here to describe the bones of the skeleton of certain species.

Sagittal crest. A raised bony ridge running along the uppermost cranium from the front to the back (i.e., in the sagittal plane) for the attachment of the chewing muscles.

Sexual dimorphism. Differences in physical characteristics in males and females. In the hominin fossil record, sexual dimorphism refers to body size, muscular attachment size, bone robusticity, and tooth size. But in modern animals this can refer to pigmentation differences, the presence of antlers, etc.

Sexual selection. Another means, besides natural selection for evolution to occur, by which individuals choose mating partners according to such things as fitness indicators.

Somatic cells. Cells that make up the body including those specific to skin, muscle, blood, and bone.

Speciation. Allopatric (speciation by way of physical, spatial separation of groups, preventing them from mating with one another).

Stature. The height of a person.

Strepsirhine. A member of the primate suborder Strepsirhini ("wet noses"), which includes lemurs and lorises. The other suborder is Haplorhini.

Taphonomy. All of the decompositional and other processes a bone or fossil undergoes between the time of death of the organism and the time until it is recovered.

Taxon (pl. Taxa). A group of organisms that share certain attributes at any taxonomic level (like kingdom, phylum, order, family, genus, and species).

Taxonomy. The classification of living things into natural orders according to their relatedness and similarities. The science of classification.

Terrestrial. Ground-dwelling or adapted for moving about on the ground.

Tetrapoda. Class of animals within the vertebrates that includes four-legged walkers (e.g., crocodiles, monkeys, lizards, horses) and those with four-legged walkers in their ancestry (e.g., birds, humans, snakes, whales).

Thorax. The torso or trunk of the body.

Trophic pyramid. Also known as the "food chain." It is the ecological ordering of a community of animals in the never-ending food energy cycle where higher members eat lower members and they, in turn, are preyed upon by even higher members. It takes the form of a pyramid because there are less and less animals per group as one moves up. The top of the pyramid holds the carnivores and they are the least populous in any given community.

Vertebrates. Animals of the subphylum Vertebrata that are characterized by having a backbone and include fishes, amphibians, reptiles, birds, and mammals.

SELECTED BIBLIOGRAPHY

For convenience, the primary citations for some of the most recent discoveries are included since they have not yet been incorporated into many of the most up-to-date books. Otherwise, the secondary sources that are most useful for seeking further information are listed.

Aiello, L., and Dean, C. *An Introduction to Human Evolutionary Anatomy.* London: Academic Press, 1990. The only book of its kind to describe the anatomical differences across the whole of the musculoskeletal system between humans and the other great apes with in-depth discussion of how hominin fossil anatomy is interpreted for each region.

Alley, R., et al. Climate Change 2007: The Physical Science Basis—Summary for Policymakers. Contribution of Working Group I to the Fourth Assessment Report of the Intergovernmental Panel on Climate Change. 10th Session of Working Group I of the IPCC, Paris, February 2007. This document contains up-to-date analyses and predictions dealing with world climate change and global warming issues.

Bahn, P.G. *Journey through the Ice Age.* Berkeley: University of California Press, 2001. Explores the oldest art with terrific photographs.

Beard, K.C. *The Hunt for the Dawn Monkey: Unearthing the Origins of Monkeys, Apes and Humans.* Berkeley: University of California Press, 2004. A great popular account of searching for fossils of the earliest monkeys and apes and how that ties in with human origins science.

Borges, J.L. *Other Inquisitions (1937–1952).* Translated by R. L. Simms. Austin: University of Texas Press, 1964. The source for the creative classification of animals in an ancient Chinese encyclopedia is mentioned here.

Brown, P., et al. A new small-bodied hominin from the Late Pleistocene of Flores, Indonesia. *Nature* 431 (2004): 1055–1061. The announcement of the discovery of the so-called "hobbits."

Brown, W.M., et al. Dance reveals symmetry especially in young men. *Nature* 438 (2005): 1148–1150. A study that links dancing ability and body symmetry as fitness indicators.

Brunet, M., et al. A new hominid from the Upper Miocene of Chad, Central Africa. *Nature* 8 (2002): 145–151. The announcement of the *Sahelanthropus* discovery.

———. New material of the earliest hominid from the Upper Miocene of Chad. *Nature* 434 (2005): 752–755.

Buss, D. *The Handbook of Evolutionary Psychology.* Hoboken, NJ: Wiley, 2005.

Calvin, W.H. *A Brief History of the Mind.* Oxford: Oxford University Press, 2004.

———. *The Throwing Madonna.* New York: McGraw-Hill, 1983. A collection of essays on the human brain's evolution and complexity including the hypothesis that the development of overarm accurate throwing played a role in the development of the large brain and language.

Cann, R.L., et al. Mitochondrial DNA and human evolution. *Nature* 325 (1987): 31–36. The "birth" of mitochondrial "Eve."

Chase, P.G. *The Emergence of Culture: The Evolution of a Uniquely Human Way of Life.* New York: Springer, 2006.

Conroy, G.C. *Primate Evolution.* New York: W.W. Norton & Company, 1990. The ultimate source for synthesizing evidence of primate evolution from the primate fossil record.

———. *Reconstructing Human Origins,* 2nd ed. New York: W.W. Norton & Company, 2005. An authoritative volume on paleoanthropology for anyone with more than an introductory interest in human evolution.

Darwin, C. *The Descent of Man.* London: Murray, 1871.

———. *On the Origin of Species by Means of Natural Selection.* London: Murray, 1859.

Dawkins, R. *The Ancestor's Tale.* Boston, MA: Mariner, 2004. A compelling view of organismal evolution taking the opposite path of tradition and going from the present to the past.

de Menocal, P.G. Plio-Pleistocene African climate. *Science* 270 (1995): 53–59.

Dunbar, R. *Grooming, Gossip and the Evolution of Language.* Cambridge, MA: Harvard University Press, 1996.

Dunsworth, H.M., et al. Throwing and bipedalism: a new look at an old idea. In J.L. Franzen, et al. (eds.), *Upright Walking.* Frankfurt: Senckenberg Institute, 2003, pp. 105–110. Reviews the hypotheses for the origin of throwing and investigates the effect of arm length evolution on the throwing ability of early hominins.

Enard, W., et al. Molecular evolution of FOXP2, a gene involved in speech and language. *Nature* 418 (2002): 869–872.

Eppinger, M., et al. Who ate whom? Adaptive *Helicobacter* genomic changes that accompanied a host jump from early humans to large felines. *PLoS Genetics* 2(7) (2006): e120.

Eveleth, P.B., and Tanner, J.M. *Worldwide Variation in Human Growth,* 2nd ed. Cambridge: Cambridge University Press, 1990.

Falk, D. *Braindance: New Discoveries about Human Origins and Brain Evolution.* Gainesville: University Press of Florida, 2004.

———. *Primate Diversity.* New York: W.W. Norton & Company, 2000.

Fisher, H.E. *Anatomy of Love.* New York: W.W. Norton & Company, 1992. A compelling look at the human reproductive strategy.

Fleagle, J. *Primate Adaptation and Evolution.* New York: Academic Press, 1999. A compendium on the biology, ecology, and evolution of the Order Primates.

Freeman, S., and Herron, J.C. *Evolutionary Analysis.* Upper Saddle River, NJ: Prentice-Hall, Inc., 2001.

Gathogo, P.N., and Brown, F.H. Revised stratigraphy of Area 123, Koobi Fora, Kenya and new age estimates of its fossil mammals, including hominins. *Journal of*

Human Evolution 51(5) (2006): 471–479. New dates that say the contentious early hominin skulls that show so much variation (KNM-ER 1470 and KNM-ER 1813) may not be contemporaneous and may actually be 250 Kya apart, so they may no longer pose a taxonomic problem.

Gaulin, S.J., and McBurney, D. *Psychology: An Evolutionary Approach.* Upper Saddle River, NJ: Prentice Hall, 2000.

Gibbons, A. *The First Human.* New York: Doubleday, 2006. The latest account of the race to discover the earliest hominins.

Goodall, J.G. *The Chimpanzees of Gombe.* Cambridge, MA: Belknap Press, 1986.

Green, R.E., et al. Analysis of one million base pairs of Neanderthal DNA. *Nature* 444 (2006): 330–336.

Hartwig, W.C. *The Primate Fossil Record.* Cambridge: Cambridge University Press, 2002. The most up-to-date book to contain accounts of the history of discovery and debate as well as the interpretations of basically the entire record of primate and human fossils.

Hawkes, K., et al. Grandmother, menopause and the evolution of human life histories. *PNAS* 95(3) (1998): 1336–1339.

Hoberg, E.P., et al. Out of Africa: Origins of the *Taenia* tapeworms in humans. *Proceedings of the Royal Society B: Biological Sciences* 268 (1469) (2001): 781–787.

Huxley, T.H. *Evidence as to Man's Place in Nature.* London: D. Appleton and Company, 1863.

Jablonski, N.G. *Skin: A Natural History.* Berkeley: University of California Press, 2006. Most things dealing with human skin and its evolution are covered. Of particular importance is the emphasis on human skin color evolution.

Jobling, M.A., et al. *Human Evolutionary Genetics: Origins, Peoples and Disease.* New York: Garland Science Publishing, 2004. A must-read textbook on the topic.

Johanson, D.C., and Edey, M. *Lucy.* New York: Simon & Schuster, 1981. A thrilling first-person account of the famous discovery.

Johanson, D.C., and Edgar, B. *From Lucy to Language.* New York: Simon & Schuster, 1996. Contains exquisite life-size or near life-size photographs of key specimens in the fossil and archaeological records for the prehistory of humans.

Jones, S., et al. *Cambridge Encyclopedia of Human Evolution.* Cambridge: Cambridge University Press, 1994. Many important discoveries have been made since this book was published, but it is a decent amalgamation of the entire gamut of research that pertains to understanding human evolution.

Kittler, R., et al. Molecular Evolution of *Pediculus humanus* and the origin of clothing. *Current Biology* 13 (2003): 1414–1417.

Klein, R. *The Human Career.* Chicago: University of Chicago Press, 1999. A great resource on human evolution especially cultural evolution.

Kuhn, S.L., and Stiner, M.C. What's a mother to do? The division of labor among Neanderthals and modern humans in Eurasia. *Current Anthropology* 47(6) (2006): 953–980.

Leakey, R.E.F., and Lewin, R. *Origins Reconsidered: In Search of What Makes Us Human.* New York: Anchor, 1993.

Lewin R. 1997. *Bones of Contention*, 2nd ed. Chicago: University of Chicago Press, 1997. A look at some of the major debates between paleoanthropologists in the present and the past.

Lieberman, D.E., et al. *Interpreting the Past: Essays on Human, Primate, and Mammal Evolution.* Boston, MA: Brill Academic Publishers, 2005. In spite of the very technical anatomical details, Chapter 7, "The Last Common Ancestor of Apes and Humans" by Peter Andrews and Terry Harrison, is especially helpful and is the most up-to-date offering of its kind.

Linz, B., et al. An African origin for the intimate association between humans and *Helicobacter pylori. Nature* 445 (2007): 915–918.

Mayor, A. *The First Fossil Hunters: Paleontology in Greek and Roman Times.* Princeton, NJ: Princeton University Press, 2000. All about how the ancient peoples of Greece and Rome interpreted fossils.

McBrearty, S., and Jablonski, N.G. First fossil chimpanzee. *Nature* 437 (2005): 105–108.

McDougall, I., et al. Stratigraphic placement and age of modern humans from Kibish, Ethiopia. *Nature* 433 (2005): 733–736. Newly redated sediments put the Omo I and II skeletons at about 200,000 years ago, making them the oldest modern human fossils on record and also putting them close to genetic estimates of human origins.

McGrew, W.C., et al. *Great Ape Societies.* Cambridge: Cambridge University Press, 1996.

Miller, G. F. *The Mating Mind.* New York: Doubleday, 2000. An intriguing hypothesis that much of the wonders of the human mind evolved by sexual selection as, more or less, courting devices.

Molnar, S. *Human Variation,* 5th ed. Upper Saddle River, NJ: Prentice Hall, 2003. The premier textbook on the topic.

Morell, V. *Ancestral Passions.* New York: Simon & Schuster, 1995. The story of the Leakey family's quest for human origins.

Morris, S.C, and Gould, S.J. Showdown on the Burgess Shale. *Natural History Magazine* 107(10) (1998): 48–55. Simon Conway Morris wrote a challenge and Stephen Jay Gould wrote a reply dealing with the issue of the inevitable (or not) outcomes of evolution.

Moser, S. *Ancestral Images.* Ithaca, NY: Cornell University Press, 1989. A fascinating trek through the history of the iconography of human origins and evolution.

Moyà-Solà, S., et al. *Pierolapithecus catalaunicus,* a new Middle Miocene great ape from Spain. *Science* 306 (2004): 1339–1344. A well-preserved partial skeleton of a possible great ape ancestor is first reported here.

Newell, M. Was it possible that Neanderthals discovered America? *Archaeology Magazine* March/April, 1993.

Nielsen, R., et al. A scan for positively selected genes in the genomes of humans and chimpanzees. *PLoS Biology* 3(6) (2005): e170.

Noonan, J.P., et al. Sequencing and analysis of Neanderthal genomic DNA. *Science* 314 (2006): 1113–1118.

Pagel, M. *Oxford Encyclopedia of Evolution.* Oxford: Oxford University Press, 2002. Entries on nearly everything related to evolution are written by the experts in their fields.

Parker, S.T., et al. *The Mentalities of Gorillas and Orangutans.* Cambridge: Cambridge University Press, 1999.

Pinker, S. *How the Mind Works.* New York: W.W. Norton & Company, 1997.

———. *The Language Instinct.* New York: Harper Perennial, 1994.

Pollard, T.D., and Earnshaw, W.C. *Cell Biology.* Philadelphia, PA: Saunders, 2002.

Portmann, A. *A Zoologist Looks at Humankind.* New York: Columbia University Press, 1990.

Pruetz, J.D., and Bertolani, P. Savanna chimpanzees, *Pan troglodytes verus,* hunt with tools. *Current Biology* 17 (2007): 412–417. This is the first account of chimpanzees making and using wooden spears. They also used them to hunt other primates.

Rak, Y. *The Australopithecine Face.* New York: Academic Press, 1983.

Reader, J. *Missing Links.* Boston, MA: Little, Brown and Company, 1981. Tales of some of the biggest controversies throughout the course of paleoanthropological science.

Rightmire, G.P. *The Evolution of Homo erectus: Comparative Anatomical Studies of an Extinct Human Species.* Cambridge: Cambridge University Press, 1990.

Rosas, A., et al. Paleobiology and comparative morphology of a late Neanderthal sample from El Sidrón, Asturias, Spain. *PNAS* 103(51) (2006): 19266–19271.

Rowe, N. *The Pictorial Guide to the Living Primates.* Charlestown, Rhode Island: Pagonias Press, 1996. The absolute best resource on living primates.

Schick, K. D., and Toth, N. *Making Silent Stones Speak: Human Evolution and the Dawn of Technology.* New York: Simon & Schuster, 1993. Includes accounts of apes trained to use and make tools.

Schwartz, J. H. and Tattersall, I. *The Human Fossil Record.* New York: Wiley-Liss, 2002–2005. Four volumes cataloging most of the fossil record for human evolution with photos and descriptions.

Shipman, P. *The Man Who Found the Missing Link.* New York: Simon and Schuster, 2001. A historical account brought to life of Eugene Dubois and his quest for fossil ape-men.

Sponheimer, M., et al. Isotopic evidence for dietary variability in the early hominin *Paranthropus robustus. Science* 314(5801) (2006): 980–982. Just one publication from the leading scientist in fossil hominin isotopic analysis.

Stanford, C., et al. *Biological Anthropology: A Natural History of Humankind.* Upper Saddle River, NJ: Pearson/Prentice Hall, 2006. A thorough, easy-to-read, and up-to-date textbook.

Stanford, C., and Bunn, H.T. *Meat Eating and Human Evolution.* Oxford: Oxford University Press, 2006.

Stauffer, R., et al. Human and ape molecular clocks and constraints on paleontological hypotheses. *Journal of Heredity* 92 (2001): 469–474.

Strier, K.B. *Primate Behavioral Ecology,* 3rd ed. Boston, MA: Pearson, 2007.

Stringer, C., and Andrews, P. *The Complete World of Human Evolution.* London: Thames & Hudson, 2005. An encyclopedia of human evolution boasting 180 color illustrations.

Tattersall, I. *Becoming Human: Evolution and Human Uniqueness.* San Diego, CA: Harcourt Brace & Company, 1998.

Tishkoff, S., et al. Convergent adaptation of human lactase persistence in Africa and Europe. *Nature Genetics* 39(1) (2007): 31–40.

Trinkaus, E., and Shipman, P. *The Neandertals.* New York: Knopf, 1993.

Trivers, R.L. The evolution of reciprocal altruism. *Quarterly Review of Biology.* 46 (1971): 35–57.

Ungar, P. *Evolution of the Human Diet: The Known, the Unknown, and the Unknowable.* Oxford: Oxford University Press, 2006.

van Dam, J.A., et al. Long-period astronomical forcing of mammal turnover. *Nature* 443(2006): 687–691. All about mammalian extinction cycles.

Voight, B.F., et al. A map of recent positive selection in the human genome. *PLoS Biology* 4(3) (2006): e72.

Walker, A., and Leakey, R.E.F. *The Nariokotome Homo erectus Skeleton.* Cambridge, MA: Harvard University Press, 1993. A scientific monograph by a large collaborative team that performed in-depth research on one of the most complete and well-preserved hominin fossils on record.

Walker, A., and Shipman, P. *The Ape in the Tree.* Cambridge, MA: Belknap Press, 2005. The story of the discovery and analysis of several *Proconsul* fossils on Rusinga Island, Kenya.

Walker, A., and Shipman, P. *The Wisdom of the Bones.* New York: Vintage, 1996. The story of the discovery and analysis of the Nariokotome *Homo erectus* skeleton.

Weiner, J.S. *The Piltdown Forgery: Fiftieth Anniversary Edition,* with a new introduction and afterword by Chris Stringer. London: Oxford University Press, 2003.

White, T.D. *Human Osteology,* 2nd ed. San Diego, CA: Academic Press, 2000. The ultimate source for bony human anatomy.

Willis, D. *The Hominid Gang.* New York: Viking, 1989.

Wolpoff, M. *Paleoanthropology,* 2nd ed. Boston, MA: McGraw-Hill, 1999. Synthesizes detailed information, particularly anatomical evidence, regarding human evolution.

Wood, B.A., and Collard, M. The human genus. *Science* 284 (1999): 65–71. A scientific evaluation of the genus *Homo* and how to recognize the earliest members in the fossil record.

Wood, J.W., et al. The Evolution of Menopause by Antagonistic Pleiotropy. Working Paper 01–04. Center for Studies in Demography & Ecology, University of Washington, 2001.

Worm, B., et al. Impacts of biodiversity loss on ocean ecosystem services. *Science* 314 (2006): 787–790. A warning that the animals that make up the seafood in our diets are on the verge of collapse.

Yoshiura, K., et al. A SNP in the *ABCC11* gene is the determinant of human earwax type. *Nature Genetics*: 38(3) (2006): 324–330.

Zhanga, J. Evolution of the human *ASPM* gene, a major determinant of brain size. *Genetics* 165 (2003): 2063–2070.

Zollikofer, C.P.E., et al. Virtual cranial reconstruction of *Sahelanthropus tchadensis. Nature* 434 (2005): 755–759. The drawing of the *Sahelanthropus* skull included in this book is based on this reconstruction.

Index

About the Author

HOLLY M. DUNSWORTH is a Postdoctoral Researcher in the Department of Anthropology at the Pennsylvania State University. When not collecting fossils in Kenya or the Republic of Georgia, she uses sophisticated imaging technology to look at their inner structures. By interpreting clues from the outer and inner anatomy of bones, she reconstructs the biology and behavior of fossil monkeys, apes, and hominins to better understand how humans arrived at their present state.